When Science
Meets Religion

L L · 9 7

An Axiomatic Unified Ontology

New Lime

· PUBLISHING ·

Patrick D. Harrington

Dedications

Firstly, this book is dedicated to my three sons, Justin, Neil and Owen and I hope and pray that this book will enable their future security.

Secondly, this book is dedicated to the memory of my father, William L. Harrington. Without him, obviously, I wouldn't exist, but, without his example of mental and spiritual openness, I would not have learned the importance of having an open mind and would not have learned how to think 'outside the box'. Also, his example of how to live one's personal philosophy has left an indelible imprint on my life and, perchance, through this book, on millions of others, as well.

Thirdly, this book is dedicated to the memories of Stephen D. Harvey and David A. Moody; for, without some of their keen insights and concepts, this work would never have been possible.

Finally, this book is dedicated to humanity as a whole; because it is humanity that deserves to know the truths contained in this book and to be able to benefit from it. It is my sincerest hope that humanity will change its view of the universe and our role in it based on the ideas presented herein and that the entire species will benefit from it and, God willing, reach the next level of our spiritual evolution.

Acknowledgements

There are so many individuals I have to thank that I hope I haven't forgotten anyone. There have been those who have contributed to the material concepts of this book and those who have helped in other supportive ways; be that financial, emotional, physical, psychological, or in some other fashion. Please forgive me if I have left you off the list. In truth, by my own theories outlined below, I ought to thank everyone I have ever met, as my life would not have been the same without them and that is a purely logical fact; but, I cannot list them all, as that in itself, would take a book of considerable size.

Naturally, I would like to thank my family for sticking by me all these years; so, a very big 'Thank You' to my mother, Marilyn Harrington nee McEwen; my sister, Kathy Wilson; my nephew, Mike Bennett; my niece, Kristi Wilson; and my brother-in-law, Dan Wilson. Thanks ever so much for always being there - for me and for each other!!

I would like to thank Richard Rapp and his wife Betty Hauser for their constancy in friendship over the years and for the countless hours and days of discussions we've had - Betty, in particular, for ensuring that I always maintain and purport a genderless God and that is surely the truth (I should also thank Betty's daughters: dear, sweet Libby and my very lovely god-daughter, Nicole, for those years when they knew they could tell me anything … you still can). I also must thank Steven Lanham (it was a pleasure being 'Best Man' in your wedding and many of the things I learned from you are written in the pages below. It was a real treat re-meeting your son, Steven; he's a true 'chip off the old block' - all the brains and more years left than we do!), and his wife Leann. Adam Murray (the heart and soul of all things dwarfish -

truly a finer friend one couldn't have) and Mark Fredericks for much the same reasons; the hours and days of time spent in their counsel has lent me countless ideas on which I've built and their friendship is invaluable. Plus, they keep my feet on the ground. Also, thanks must go to Mark's wife, Debra Penna-Fredericks, for teaching Mark to value life in ways he might never have known without her. Nor must I forget Mark Talkington for teaching me the ways of the trees. Special thanks to all of these people, for without them, my life would have been so empty. And we learned together that there are eight ways of looking at things.

It would be most unkind to not thank the several members of the Google Group "Minds Eye", who have been enormously helpful to me over the past four years by acting as a soundboard for my concepts and, certainly, as worthy opponents with which to argue - whether you agreed with me or disagreed. So, I give a very big 'Thank You' to Craig (the founder of the group), Neil Terry, Chris Jenkins, Molly Brogan, Juan Montoya, Ian Pollard, Lee Douglas, Allan H., Gabbydott, Rigsy, Douglas Blight, the Ornamental Mind of Bill Faust, E. A. Doherty, Ash Kashal, Manfraco Laws, Paradox, Slip Disc, Chazwin and Pottsie and, indeed, to any others I may not have specified.

A very special 'Thank You' and huge acknowledgement of her material contribution to the barrister, Lucy Tapper, who asked me the question "What about Emotion?", which resulted in the chapters entitled: *What about Emotion?* and *How Does This Change Justice?* It's amazing how just a three word question led to a 5,602 word answer, eh?

Indeed, I would be very remiss not to thank Des Knox for being one of the best friends I've ever had. Plus, he has, for the past 3 years, acted as one of my very first sounding boards for any new ideas that have sprung up. Without his feedback, this work would not have reached a point of acceptability as far as I'm concerned; so, thanks, Des!

Also, I'd like to thank the several women, who, throughout the years, have been an inspiration to me - whether or not they knew it: Sarah Atkinson - nee Wright (the mother of my three sons); Mona Dashney (although I knew her as Mona Halliburton at the time. I couldn't be happier that you've found happiness with Mr. Dashney and, of course, congratulations on becoming a grandmother!); Kathy Russell (my 'High School sweetheart'); Letitia Kannes Ruth Martiníque (for being a guiding light in the darkest of nights); Jane Wagnon (no, I haven't forgotten you. I paid far more attention to you in 3rd Grade than I did to our teacher, Mrs. Rosenthal); Sherry Huffman (yes, it WAS 5th grade, and, of course, I could never forget you); Linda VanDyke (a truly remarkable individual: beautiful, intelligent and extremely musically talented. I was SO glad to read that you've become a recorded artist! I always knew you'd do well. Yes, that does mean I 'Googled' you; I gladly admit it. And, whether or not you admit it, you DID kiss me on the lips one recess in 2nd Grade [the highlight of my life at ages 7-8!] and Doug Backlund will always verify that for me—sufficiently embarrassed? Good. ;-)) Karla Conditt (for being the epitome of honour, class and style); Christina Thurmond (a fabulous friend and a very funny and classy actress—hire her for anything from slapstick to Shakespeare!); Bridget Walker (for your hours and days of counsel and friendship); Pietrina Nicastro (for being beautiful in every way; here's the proof that you were right in thinking I enjoy raising people's eyebrows); Laura Dearduff (for always being there. I will always miss you.); Sylvia Barron (not just for being a great friend to me but the best friend my sister could ever have); Aerwana Dearduff, Mary Ewersmann, Cathy Gladden, Rita Hensley, Sandy Scott, Barb Barron, Patty Caputo, Cathy Huddleston, Vicky Hunter, Carla McGuire, Bonnie Sherman, Jodi Walker, Karen Anthony, Diane Bargmann, Kristi Bayles, Tracy Bromberg, Kim Burgoyne, Josie Calloway, Joan Chandler, Judy Copeland, Sharon Dively, Linda Farasy, Denise Fogarty, Mary Godfrey, Corless Hentz, Lois Hamil, Debbie Hill, Sue Huddleston,

Karen Jarmusziewcz, Kathy Johnson, Angie Lochmann, Connie Lohmeyer, Yvonne Macaré, Sandy McCullough, Shari Millikan, Regina Morales, Cindy O'Brien, Lisa Parker, Catherine (Casey) Plannette, Melinda (Mimi) Pomeroy, Janet Quick, Norma Rhoads, Veronica Ross, Yvonne Scanlon, Lisa Schaedler, Diane Settle, Sally Shenk, Ann Stade, Vici Strini, Julie Wade, Jeannie Williams, Sandy Wilson, Kim Winder, Dianna Bilyeu, Krista Bromberg, Laura Brown, Lisa Maass, Mary Quick, Sherry Stoien, Mary Tippett, Michelle Walters, Mary Wilson, Dawn Buerck, Sheila Chapman, Bridgett Clarke, Donna Ferrari, Dee Dee Griggs, Mary Hastings, Andi Midkiff, Michele Nicolay, Jeanna Olson, Louise Riney, Fiona Chater, Gemma Yates and Susan Hagan (for brightening up my days just by being there).

Last but not least, the men-folk who have been inspirational to me in a variety of ways: Dave Baynes (it's a shame we lost touch with each other. You were always the 'Starsky' to my 'Hutch'.); Larry Brooks (a great friend who moved too far away far too soon); Richard Ehrhard (I'll never forget you, mate. You and Dave Baynes and I shared so many great times together.); Dave Graves (one of the more solid friends a guy could ever have. I'm so glad I was able to introduce you to your wife, Debbie [nee Wall], who has also been one of the best friends I've ever had and a quick "Hi" to the kids Bryan and Chris); Gary Hagan (a true 'diamond geezer' who never let me down as a friend. One of these days, we'll have to catch up, again, as I missed you on my last trip back to St. Louis and that was something I didn't want to have had happen.); John Hagan (You're like a true little brother. Well, you are to Gary and an older brother to Susan; but you're still like a "little bro'" to me, too.); Steve Klenk (Hats off to you, mate - the snow white Duke of Prussia! I say that, purely, because I almost never saw you without your fedora... in true 'Indiana Jones' style!); William 'Clacy' Loveless (a close friend and one who saved my back in a 6-on-1 situation, making it 3-on-2 - I'll never forget you for that!); Jim 'Jimmy Otto' Schuermann (an incredibly great friend and, without doubt, the most intelligent person I've ever met.); Richard Sharp (a friend, confidante and housemate that I will never forget - especially your 'Ozzy' impersonation. I hope all is well with you!); Ed Sykes (one of my oldest [although you're younger than I am] and dearest friends. I was so glad to see you finally found your dream-girl - and you couldn't have done better even if you went all the way to Canada - oh, that's right, you did!); Bob Weiss (a gem of a guy, without whom I would never have become interested in playing the drums. You left us far too early, Bob, and I'm sure your brothers, Don and Steve, would heartily agree.); Bob Davis, Jimmy Davidson, Mike Ferrante, Ed Peebles, Perry Rolfingsmeyer, Matt Sudbeck, Geoff Acton, Rick Black, Don Blake, Tim Gorman, Rick McDermott, Charlie Orso, Doug Backlund, Craig Book, Earl (Eugene) Brizendine, Lloyd Bryan, Randy Bunk, Charlie Butler, Pat Cody, Quirt (Lonnie) Collins, Mark Crosby, Dave Ewersmann, Pat Fogarty, Dave Gerner, Mark Goss, Chris Greiling, Steve Hargrove, Randy Hart, Kevin Haynes, Rick Hipes, Scott Hutchison, Keith Iborg, Derek Irving, Kurt 'Coach' Jacob, Keith Jeske, George Johnson, Vernel 'Little Joe' Johnson, Jeffery Keen, Dan King, Doug Kline, Dan Kroll, Aurelio Lee, Fred Lee, John Lenhardt, Mike Lorenson, Jim McCullough, Dave McDaniel, Mickey Mink, Fred McGee, Jometric McIntyre, Charles 'Mike' Meisch, Bob Meives, Steve Mercer, Scott Miller, Norman Montgomery, Jeff Osborn, Tim Perkins, Dave Pilla, Eric Ritter, Dave Scherer, Ed Schmermund, Dan Scott, Dave Soest, Rob Stewart, Dave Suchanek, Kent Thoroughman, Howard Tweet, Steve Vancil, Joe Vetrone, Jeff Walters, Ken Ward, Doug Williams, Dan Zarlenga, Bill Hostmeyer, Tom James, Mike Kolley, Art Lum, Jim Marshall, Anthony Montgomery, Terry Orr, John Remstedt, Chris Rootz, Paul Shepherd, Scott Spangenberg, Bob Walters, Bob 'Bear' Baranowski, Doug Fogarty, Scott Janssen, Harry Linnenbom, Mike Marrah and Jim Viehman (for rounding out my life and making it worthwhile).

Contents

About the Author

Patrick D. Harrington was born on 3rd April, 1963 in St. Louis, Missouri, USA. He has, over the years, studied most aspects of science such as biology, chemistry, physics, geology, oceanography, psychology, medicine and astronomy. He has also studied various religions including Judaism, Christianity, Islam, Hinduism, Buddhism, Taoism, Confucianism, Sikhism, Voodoo and Shinto and has, in the cases of Judaism, Islam and Buddhism, heavily researched their more mystic aspects. He has also studied the more ancient faiths of the Mesopotamians, Romans, Greeks, Egyptians, Celts, Norse and the pre-Buddhist era Tibetan Bon faith, the Australian Aboriginal faith, as well as some aspects of lesser known African animist beliefs. He is fully conversant with the beliefs and practices of modern Paganism and Wicca and their Celtic and Norse origins.

He studied Spanish and German formally and taught himself to read Hebrew and Greek and is currently teaching himself Arabic. The latter three were studied in order that he could read the original Scriptures of the Old and New Testaments of the Bible and, obviously, the Qur'an, as he did not wish to be influenced by the agenda of translators nor have to rely solely on translations of these Scriptures. He has also familiarised himself with a multitude of other languages and, because of those studies, is knowledgeable of the science of linguistics and the development of languages over time. Some of the other languages with which he has familiarised himself to a lesser extent include: French, Italian, Latin, Russian, Polish, Welsh, Japanese, Hindi/Urdu, Aramaic, Ancient Egyptian and the Assyrio-Sumerian languages.

In addition to these scientific, linguistic and religious studies, he has studied the philosophies of Socrates, Plato,

Aristotle, Saint Augustine, Saint Anselm and Descartes. These studies have given him a broad base and an enormous wealth from which to gather information and to synthesise that into the following work. Perhaps the greatest sources from which this synthesis derived were Plato, Saint Anselm, Rene Descartes and from Professors Albert Einstein, Steven Weinberg and Brian Greene.

Irrespective of the broad background above and far from being a 'professional student', he has spent most of his working years employed in IT roles as a computer operator, database administrator and computer programmer. With regard to computer programming, he is also self-taught; but, he has never found that a barrier to employment. Being employed as a computer programmer for 12 years forced him to use formal logic on a daily basis and it is that, which has proven to be the crucial key to seeing the logical relationships behind certain philosophies and the workings of the universe, which has led the author to the realisations explored in this book.

Author's Note

What lies after this note is the end result of over 5 and a half years of the project that became this book: the project of trying to put together all the pieces of the puzzle that I'd accumulated up to that point. What puzzle; you may ask? It is the puzzle that is our universe. Most likely, although I have found answers to questions no one else has, I have barely scratched the surface of truth. Finding the pieces was the easier part; putting them together into a whole that made sense was far more difficult, indeed. And this was no ordinary puzzle; there was no picture on the outside of the box that showed what the result should look like. That, too, was a mystery. Whilst I feel I have done a reasonable job, I don't, for a minute, believe that the job is fully done. I still see a few unanswered questions here and there. Perhaps a follow-up book to fill those in will come later. But, at some point, I had to stop the research and compile what I had in order to present it to the public and I present it to them for their sake.

      This work began, in earnest, when I decided to try to use String Theory in order to solve the problem between the Standard Model and Quantum Entanglement, which is explained in greater detail below. The short answer was to define the length of a single dimension of the multi-dimensional, Calabi-Yau space, the extra dimensions that String Theory affords. After that ever-so-small definition, which was completely within the rules of physics, an epiphany struck me. I noticed that this new geometrical configuration had philosophical implications that were enormous, to say the least. The geometry of the system uncovered that a single monad is that which exists. All existence was, in essence, the ends of vibrating strings bent and folded around the various dimensions and that all reality was a function of one entity. The implication was that, in fact, there was a God and this single entity - this monad - was it. The various and multitudinous things that we perceive as separate are but the very ends of the extensions of this single monad only seemingly separate because of the way they are extended, bent and folded throughout the dimensions. The details are, of course, below. When I realised that God could, essentially, fall out of the equations, I was dumb-struck but I knew I was onto something and onto something big - big enough to be worth writing about.

Soon, other parts of the puzzle started falling into place and the whole picture of the universe from macrocosm to microcosm unfolded before me. Now, it's time to hand over the work so that the world can, hopefully, benefit from it. I know that I've just contradicted myself in saying above that the 'whole' picture unfolded when further above I stated that there are still unanswered questions. Hopefully, you will share my belief that those unanswered questions are small in comparison to the importance of relating the overall picture. I believe, after reading this book, you will understand the general scheme of the universe, its general plan and our role in it. Finally, I sincerely hope that you, in your individual way, will take an active role in helping the world reach the next plateau of spiritual evolution and that we as a species, at last, can grow together and become stronger for it.

I thought I should make just one last special note regarding the suspected faster-than-light neutrinos that were shot between CERN in Switzerland and Gran Sasso in Italy. Firstly, they need to be sure that their measurement of the distance is within 18 metres and their measurement of the time is within 60 nanoseconds in order to make any statement at all. This raises the question: when did they last measure the distance?   If there had been any seismic activity since their distance measurement, the distance could have changed. Secondly, there is no way to mark neutrinos as coming from a particular source. That is, you can't write a return address on them. Every star emits neutrinos and can do so in any direction at any time. This raises the question: was there a star that had emitted neutrinos in the very direction that, eventually, would have lined up with the expected destination at Gran Sasso? Such a star could be 10, 100, 1000, or millions of light-years away and the truth of the matter may be that the beam of neutrinos that was captured in Gran Sasso was from a far-away star that occurred an immeasurable time ago. Given the backdrop of billions of stars in our own galaxy, there is a reasonable likelihood that this could have been the case. The investigation to discount that could prove very difficult to perform. If they discount this possibility, then they are discounting reality. The only way to prove whether or not neutrinos can move faster than the speed of light is to be able to 'regularly' repeat the experiment given different equipment (of a similar variety) and in different sources and destinations. Personally, I think they will discover their measurements incorrect or unrepeatable and we're back to normal. The only faster-than-light-speed 'truth' lies in quantum entanglement and, below, I put forward a perfectly reasonable answer to that.

Preface

Shakespeare was quite right to have Hamlet question whether to be or not to be. For it is in the questioning of one's existence where one will find their highest calling. But only one who was given no choice BUT to exist can contemplate that continuity. The question of why we exist at all is the foundation upon which we, who consider ourselves to be intelligent, form our codes of behaviour, our ethics, our individual morals, our laws, our society and that by which we, as members of our species, define how to be. Finding our place in the universe is more than just locating our sector of the galaxy in relationship to other astronomical systems; it is defining the reason for our existence - the purpose for our intellect and the goals towards which we should direct it.

So what is it to exist? Existence comes in many forms. Ideas exist. But where? In our minds or in a Platonistic abstract plane? The things we see in dreams exist. We can interact with them and we can be emotionally affected by events in dreams. Are dreams played out solely in our minds or are they our actual exploits in a vast dreamscape, an astral plane, where some aspect of us is able to interact freely therewith? And what of our waking experiences? Are they too just played out in a large-scale virtual reality invention of our minds or are they just another aspect of a complex interplanary entity that is able to interact with physically real objects on a physical plane? No one seems to have a problem with the basic concept of a physical plane in which there exist real tangible objects.

Most people don't have a problem with associating a certain level of reality to dreams. But we "know" that dreams aren't real - well, we know they aren't physically tangible and that seems to be a pretty good indication of reality. Magnetic fields are intangible yet recognized as purely physical. If an intangible magnetic field can have an influence on physical objects, its existence as a physical reality is confirmed. But dreams can inspire changes in us that can be expressed in the physical world, so is that not akin to influencing physical objects?

Dreams can motivate us to take action in waking life. Many of humanity's nobler individuals have had dreams that moved them to pursue scientific discoveries, begin philosophical movements and concoct wonderful inventions. On the darker side, dreams have moved men to murder, war and suicide. Several stories have been inspired by dreams and people have expired physical energy to read them or see them in theatres and cinemas. And these stories have varying degrees of emotional impact on us as well and may inspire us to dream or to act differently. These are no less real effects in our universe than the alignment of metal filings in a magnetic field.

Ideas, though, are a tricky realm. You can have them either when awake or in a dream. They are accessible to us in both of those types of reality. Is it because they are solely in the domain of the mind, which we take with us in both waking life and in dreams? They are unlike memories because their existence is perceived to originate, they "occur" to us. Sometimes two people who are near to one another seem to have the same idea at the same time. Is this a spontaneous coincidence or is it because two mental fields = with a larger area of influence than we imagine - are intersecting across both our physical space-time and an abstract plane? It seems to happen more between some people than others. Is this, too, a coincidence or is it because some mental fields have a greater affinity with one another? Perhaps mental fields behave similarly to electromagnetic fields in that the mental waves that make them up can interfere with or amplify other mental fields within their area of influence. There's no evidence to suggest otherwise and the behaviour is consistent with what one might expect to see if that were the case.

Field theory has traditionally been presented as a physical phenomenon but human mentality, whether it is cognitive or emotional, seems to behave in accordance with every aspect of it. It has long been expressed, by those who claim to experience it, that telepathy involves the sending and receiving of messages via a hitherto unresolved medium through which mental images and thoughts can be transferred. The term "medium" itself stems from the belief that spirits are located in a parallel plane of existence to our own physical plane of existence and that a medium can channel spiritual energy flowing through this ethereal plane and allow them to form an interface between these realms. Is this just a fanciful claim made by ignorant, superstitious people or is there a rational and reasonable explanation that supports the possibility?

So what is it that exists? Airplanes exist. But they didn't use to. Dinosaurs existed but they don't anymore. Physical existence seems to be transient by all reckoning. Things appear and disappear. Everyone we will ever know will have either already been born or will be born before we die and everyone will either die before or after our death. There is nothing neither surprising nor shocking about that. Yet the stuff we are made of, the molecules, still remain. Or do they? No, actually, they decay or change into other molecules. But the atoms remain, don't they? Well, for a lot longer, certainly; but, in a couple of billion years or so as the sun enlarges and engulfs the Earth, these too will decay, or change into simpler forms. The only substance, if it can truly be called a substance, which remains unchanged for all time, is energy.

Energy has many forms and types and several known means of transforming from one form into another. Even in the process that physics describes as annihilation, where two particles of matter and antimatter collide, e.g., an electron and a positron, is only a transformation into light and heat, which are simply different frequencies of electromagnetic radiation. It is true that

9

the particles are annihilated but the energy contained in those forms transforms to free energy, light. More properly, it transforms into electromagnetic waves, photons, of varying frequencies, thus the light and heat (heat is only electromagnetic waves with a frequency in the infrared range). It is, then, a reasonable assumption to say that, if there is anything that exists, it is energy. Energy exists. As a mathematical assertion it can be put plainly into two symbols: $\in E$.

What else can be said about energy? In order to be able to induce a reason for (the answer to "why?") its existence, one must first define the properties it has and how it behaves. Answering the other adverbial questions (how does it exist, how much is there, where does it exist, when does it exist, what caused it to exist, in what conditions can it be found) should help. What else can be said about existence? There have been many famous ontological or cosmological theorists in the past: Plato, Shankara, Aquinas, Anselm, Spinoza and Descartes to name a few. These philosophers each had slightly different angles from which to approach the problem but they were all founded in religious beliefs. Plato divided the universe into two parallel worlds, one physical and one metaphysical. Shankara saw the world as an illusory manifestation the whole of which being one and the same with a God that he called Brahman, the only entity that really exists. Aquinas used a more cosmological argument to arrive at a theory for the existence of God as a First Cause. Anselm attacked the problem by defining God as that than which nothing greater can be conceived and then proving, with logic, that such a God must exist. Spinoza believed more in a deific God, one who sets up the reels of Creation, then sits back, and lets the film play. Descartes realised that recognition of his own existence as a thinking thing implied that there was a conceiver greater than himself that must precede himself and be so perfect as to be able to conceive and realise all of reality. Each of these men looked at the universe and saw different things. But is there a sound explanation that could account for these differing views? Energy exists, of that there can be no doubt and the fact that energy exists is the underlying axiom of this work; but, how could energy - pure, raw, unadulterated energy - be responsible for all the physical, emotional, mental and spiritual aspects of our human existence? Let us question our views of reality and challenge our long-held beliefs whether or not they are scientific or religious; but let us not doubt the truths that can be derived from either.

In this book I endeavour to pick up where Plato, Shankara, Anselm, Spinoza, Descartes and Einstein left off and give those minds the concept of string theory as we know it today together with the Standard Model and Quantum Mechanics. Let these mindsets, as I perceive them (for, in fairness, I can do no other), mull over the facts along with the Scriptures and Holy Writings of those faiths that have them and answer the questions of existence and our role in it. Then present the result of those combined mindsets and put forth a vision of reality as it most likely is. I fully share the experience of standing on the shoulders of those giants. For example, it is the mindset of Spinoza that tackles the question of 'creation ex nihilo' below. Without the previous work done by these intellectual giants, I could never have been able to put it all together into what I hope you will accept. Some of it will be hard to accept; but this is not because it is untrue or hard to believe, but that it rocks the foundations of our view on life and our role in it.

Follow me on a journey through space and time, a journey that will take you to a time before the Big Bang, lead you through the quagmires of ontological arguments, guide you through the intricate causeways of String Theory and land you, finally, in the omni-reliable hand of God Almighty. Science and religion have not always seen eye-to-eye throughout history but see what happens when the two are woven into one holistic tapestry made of the most precious thread that has ever existed: Energy.

Permitte Phoenix Reviviscere et Permitte Saeculum Novum Incipere - Let the Phoenix rise again and let the New Age begin!

When Science Meets Religion

PART 1
Cosmology

What is Truth?

What is truth? Is that cosmological? This is the first chapter of a part of a book entitled Cosmology and it begins with asking 'What is truth?' Yes, it is, in essence, a cosmological question. Scientific truths like those regarding the formation of our universe surely occurred, but how and why? Recently, Prof. Stephen Hawking made the statement that 'Science has made God redundant' in his book *The Grand Design*. This is certainly not even a fair scientific statement; although it is, obviously, part of *The Grand Design* to make certain incredibly respectable men of knowledge state that that is the case. Plainly and logically, if it is true that God exists as an omnipresent, omnipotent and omniscient entity, then science could not possibly make God redundant, as, by definition, such an eternal entity could never be supplanted; and, if it is true that God does not exist, then science has not caused that to be the case and, therefore, has not made 'that which never existed' redundant. It's simple-minded, illogical foolishness that thoughts such as these enter the minds of great thinkers and, quite honestly, I can hardly believe it possible that those gifted with such minds would think such thoughts. That's not how science works and scientists of Prof. Hawking's calibre ought to know better than to make such nonsensical statements, unless it is simply done to provoke rather than to provoke thought. Whilst I, too, am at times happy to provoke; I do not provoke without the deepest of thought regarding the provocation and no provocation of mine will be found to be baseless.

Science gives us the answers to 'how' and religion gives us the answers to 'why'; this is how (and why!) God answers the questions science cannot. You see, not all of the 'whys' can be answered without admitting we have purpose - this is a teleological aspect, which I will show to be imperative and scientifically inescapable. There must be a truth behind our existence that is strictly ontological to the quantum level of detail; yet, that which exists must have come into being or appearing as it does now, and that is cosmological. But how and most of all… why? Is it true that if you know why, then 'how' doesn't matter as much? Well, yes; but, that is no excuse for improper investigation that leads to discovering 'how'.

This, "What is truth", was also a question asked by the Roman Governor of Judea, Pontius Pilate, to Jesus during Jesus' 'trial' before Pilate. Pilate, being a politician, understood that there are different kinds of truth in addition to the fact that some people accept certain arguable concepts as true that others do not. He knew also that, sometimes, the truth itself is less important than what people believe. A people that believe a falsehood or fail to believe the truth fool themselves. Pilate wanted to discover what Jesus understood as truth. The question Pilate and Jesus were discussing was whether or not Jesus was a king of some sort. Jesus stated that he was a king but not a king of this world and it was at this point that Pilate asked the question, "What is truth?" and, ultimately, said he found 'no fault' in Jesus and would have dropped any charges against him. The High Priest would not let that happened and the rest, as they say, is history - disputed, naturally. In the Gospel of John, where this discussion is elucidated the fullest, Pilate's question was never answered. Neither was it answered in any of the other Gospels. This doesn't surprise me as the two parties would have had enough difficulty in communicating at all much less discussing the intricacies of various philosophies regarding truth. Pilate, who spoke Latin and Greek, knew no Aramaic and Jesus who knew Aramaic, Hebrew and Greek, knew no Latin; so, they must have conversed in Greek, as that was the tongue they both knew and it was the 'lingua Franca' of the time thanks to Alexander's massive impact on the ancient world.   So, the question remains unanswered in the Gospels. Shame. It would have been handy.

I believe a discussion of truth at the start of this work is vital, as there are many kinds of truth and there are those who only regard one variety as valid. Of course, I'm referring to the world of Science (with a capital 'S' to denote the institution and no one in particular) and the concept of a scientific truth. A scientific truth is a truth that can be demonstrated by repeated tests, each of which will produce the same result. Sometimes, a scientific truth can be demonstrated, repeatedly, by testing under varying conditions and sometimes only under certain controlled environments. Nevertheless, it can be demonstrated repeatedly and the result will always be the same, thus deserving to be called truth.

But there are, as I've said, other forms of truth that cannot be repeated. The simplest example of this is an historical truth. For example, there was a cause for World War II. What this cause was would be an historical truth. Some people might say that it was when Germany invaded Poland, as this caused Britain to declare war on Germany as it had signed a treaty, earlier, with Poland stating that they would defend Poland if they were attacked. But, could it be then, that the treaty itself was the cause? Without the treaty, Britain would have had no reason to enter the fray. But we can't go back in time and test our theories because we can't go back in time. So, whatever the true cause for World War II was, it cannot be a scientific truth because it cannot be tested or repeated - not under the same circumstances nor under different circumstances. It is an historical truth; but the cause of World War II is no less a truth than any scientific truth; it is simply a different type of truth.

There are emotional truths, as well. For example, given a particularly sad film to watch, some people will feel compelled to cry because a particular scene (a given stimulus) causes a reaction within the person that causes them to emote their feelings by crying. Others, however, will not be so moved. Others still may be moved at first, but after watching it a second or third time, the stimulus no longer causes the effect. Yet there are those who will always be affected in the same manner no matter how many times they watch it. The truth of whether or not the stimulus will result in a person crying is an emotional truth and will vary from person to person. It is no less a truth than a scientific truth or an historical truth; it is simply a different type of truth.

Without going into examples of each of the following, I will state that there are moral truths, ethical truths, philosophical truths and religious truths as well as the types already discussed. Of these, religious truths are, perhaps, the most arguable of the lot and, because of that very fact, have caused (in an historically true way) many wars, battles and (in an emotionally true way) bad feelings between people. Perhaps when it comes to religious truth, it would be better if I did give an example. Now, in order to protect myself from offending anyone by doing this, I will affirm that I'm only doing this in order to give an example and not trying to stir up trouble by siding with one side or vilifying another and adding even more grief to the world; I merely want the reader to understand what I mean by a 'religious truth'.

The simplest example I can think of is that, in Christianity, it is a religious truth that Jesus, the boy raised by the Jewish carpenter Joseph of the tribe of Judah and his wife, the child's mother, Mary of the tribe of Levi, was the Son of God incarnate. Both Jews and Muslims disagree with this and Buddhists and Hindus feel that it is, largely, irrelevant. Muslims regard Jesus (called Isa in Islam but would, in truth, have answered to the name of 'Yeshua' in his native Aramaic) as a prophet and Jews view him as an heretic. These are religious truths. To those who accept only scientific truths, these 'facts' are simply opinions. But, they are more than simply opinions to the believers; otherwise, there wouldn't have been so many wars and battles waged because of differing religious truths. To be more technical about religious truths, perhaps it is fairer to deem them as accepted axioms. They

18

are concepts which must be accepted if one is a holder of that particular religious faith.

Yet Science, seemingly, has an unwritten axiom too: that God can never be the answer to a scientific truth. This unwritten yet widely accepted axiom of Science is in no way, shape or form, a scientific truth. In fact, it is, scientifically, completely baseless. Science will state that there is no evidence for God, therefore it can be dismissed. Logic, though, states clearly that one cannot make a logical and truthful conclusion based on a lack of evidence - especially if there is, truly, NO evidence, which is a profound lack of evidence. So, the dismissal of the existence of God by the world of Science is completely illogical. It would be intellectually more honest if Science held up its hands and claimed that it simply doesn't have an answer with respect to the existence of God. This would be, technically, an agnostic view, which is in perfect keeping with a logic-based scientific world-view when one doesn't know the truth. Because it takes an anti-theistic view and does so without stating it plainly, this is where Science falls over in a heap and is distrusted by those who have had personal experiences that are historical truths that offer them evidence that would lead one to accept the existence of God as a truth.

Scientists, though, cannot repeat any of these personal experiences, although they cannot explain why. Because of the subject matter, they usually don't even want to try. If, by chance, they did succeed in replicating some kind of experience that would lead one to believe and accept the existence of God, their precious unwritten rule would be exposed for the blatant, biased and illogical axiom that it is. Scratch that last, I've already done it myself. I strongly suspect that the reason that religious experiences cannot be repeated is because there is an historical component to them. They happened to a particular person at a particular place at a particular time and that spatio-temporal environment can NEVER be replicated; so, any attempt to replicate all the other aspects still fails as the place and time is different.

While many scientifically-minded people will scoff at such an answer they cannot, logically, discount the obvious importance of relative spatio-temporal coordinates in a space-time continuum that is governed by relativity!

Science also does not accept as truth any hypothesis that cannot be tested. This is, for science, a very good idea, as it means that, while sometimes science can be slow, it is always sure. Well, no. It's really just saying 'as far as we can tell'. And every day it grows and changes. Science's greatest victory of the truth is constantly changing purely because we've been wrong and learned from it. As long as we learned from it, it was science. It all started with humanity's need to know its true environment. That truth must account for all events however rare and explain them. Otherwise, it is wrong, perhaps in several ways - certainly in as many regards as there are events it cannot explain. I just want to step outside the box and see what could be put together given what I know and can see.

What I purport in this book are, scientifically, merely unproven hypotheses; but, that does not automatically mean that they are untrue. It only means that, as yet, they have not been repeatedly tested and found to be true; after all, I've only just put this together. There may come a time when some of what I state can be tested and proven or disproven either directly or indirectly, although, as most of it is based on String Theory - which is mathematically proven but not empirically proven - I doubt whether Science will accept all aspects of my model of physics; but, I believe they will find some new answers to certain questions that are both sound and falsifiable. Nevertheless, there can be no testing without first having laid down a hypothesis to test and with that in mind I see no reason to withhold my ideas simply because Science, at the moment, doesn't have the capability to falsify my work. I view this as a failure of Science more than a failure of mine; Science simply hasn't caught up with my concepts and I don't believe that should be held against me or my model of physics.

I believe that, since my model answers far more questions than either the Standard Model or Quantum Mechanics or current String Theories, it is true and is more likely to be true because it answers more questions than any other model yet conceived. As this model of physics has, latent within it, both religious and philosophical implications, it should be of interest to scientists, philosophers and religionists of all kinds - it is plainly relevant to everyone. This model has built-in metaphysics that current models completely lack and, because it has built-in metaphysics, it answers or helps to answer questions that science cannot or dares not to approach. The implications of my model lead us into the realms of philosophical truths and religious truths and I will not turn away from pointing them out. This will, of course, mean that my model will support some philosophical and religious truths and will not support others. That, again, is not my fault. If this model is correct, then the implications are philosophical and religious truths that the world should accept as they would accept any kind of truth.

In fact, the current understanding of the Standard Model in light of Einstein's Special Relativity also comes with some very important philosophical implications and, because that theory has been empirically proven, I will point them out so that the world can accept them as truth. It may come as a shock to some and will overthrow what they may feel is a religious truth, as well. This particular truth I'm discussing is the concept of 'Free Will versus Fate' and there is a chapter devoted to it. As the truth can only be one or the other, roughly half the people will feel vindicated and the other half defeated. But the truth is what matters and we should always accept the truth or learn to live with it as best as we can. Sometimes, it is the philosophical implications - the philosophical truths - of science that matter more to our day-to-day living than the science itself, as they affect our approach to life; this is one concept I truly want the reader to appreciate because my model addresses areas for which modern science simply has no answer or, if it (modern Science) does attempt an answer, the answer is a dismissive and disrespectful 'human psychosis'.

For example, some spiritual experiences like ghosts and angels that are reported by so few people and the experiences simply don't repeat, science completely dismisses them. I can't do that because I want my theory to be comprehensive. Simply because an event is rare does not mean that the person who has experienced the event is mad. THAT kind of thinking is madness. It is rare for a flipped coin to land on its edge, but I've seen it happen. Granted, it was because the carpet was a shag carpet with enough pile to hold the coin between the fibres, but, nevertheless, the rare event occurred. And, although we tried for an hour, we could never get it to happen again. My point is that I don't discount rare events nor is there any logic behind dismissing them; rather, in my model, I try to find an holistic reason that allows for them. In some particularly rare cases/events that the universe requires, you won't be able to repeat them. An example of this is the Big Bang. It had to happen to get the universe started, but, because it requires all the energy in the entire universe to exist at one small point in space-time, you don't want to have another one happen in your kitchen or front garden! The conundrum there is that it requires the kitchen and the garden as a small but vital input for the Big Bang; as you don't get a Big Bang without ALL the energy there is.

As I've stated, my model will support some views and not others and this will cause, without fail, a certain number of detractors that are alienated by the philosophical and/or religious implications of my model. This is as unavoidable today as it was for Galileo in his day - save that the Holy Roman Catholic Church has far less material and temporal power now than it did in Galileo's day. In the extreme case, some people may call for my death, as they will believe me to be an heretic or just another vile espouser of evil. Others will, hopefully, applaud my work as being ground-breaking and giving a fresh, new outlook on the universe. Still others will be completely nonplussed and there will be every possible degree of variations of reaction in between. All these reactions I've anticipated but the inevitable negative reactions will not put me off from revealing what I firmly believe to be the truth. I have complete and total conviction in my model or I wouldn't present it to the world. If I thought, for a moment, that I was wrong, I would hold back my hand. But here is the work.

I hope and pray that the readers will enjoy this fresh, new view of the universe even if it doesn't change their lives or their philosophical views. I believe that this model and the philosophical and religious implications derived from the model are actual truths. It has taken years of research (my first doubts about the veracity of certain religious truths began when I was 7 years old and, at the point of this writing, I'm 48 - that's 41 years of trying to discover/uncover the truth) to get to a point where all the pertinent areas have been covered and I think I've reached that point.

The main intention of this work is to relate truth. Whether that truth is potentially scientific or whether it is a philosophical or religious truth (either based on issues of faith or having political implications), it is truth that I intend to impart. And truth based on existence, that is, this model stems more from an ontological view than a cosmological view; nevertheless, it contains, as it must, a cosmological component, as that question must be covered by any theory that attempts to explain as much as possible; so, cosmology is where I begin. Science works with knowledge but ontology must precede it, as knowledge implies not just a knower but a knower that exists; so, it is ontology and the discovery of that which exists that is our guiding star. Is that knower simply us or are we a projection of an ideal knower who, not only knows like I do but also knows exactly like each of us? How do we uncover the truth? In order to do this, some long-held definitions of terms must be re-interpreted in order to reflect a more accurate view and those definitions need to be laid out first before I get to the meat of the argument.

The second intention of this work is to draw together science and religion and impress upon the reader that, in many cases, the two speak of the same concept but use different terms or language to refer to the concept. This difference in terminology stands in the way of understanding the truth and I will do my best to point out the areas where science and religion discuss the same topic using different terms. I also do my best to make science and religion meet and meet on friendly terms. I have never believed that the two angles were diametrically opposed, as many think they must be.

Logic dictates that they certainly are not by necessity and logic is a very good guide to the truth. My work as a computer programmer has given me a decent understanding of logic and its applications. It - logic - is one of the crutches I use most to help me walk towards the truth and stand behind the truth. I also look for analogies, as I believe that they are a very good way of demonstrating likenesses to people who may not have the technical or mathematical background to easily understand some of the intricacies of the model. I would think that no one will have a problem understanding the analogies I employ; therefore, I expect any reader should be able to fully comprehend the model and the philosophical and religious implications derived from it.

Sometimes, the truth is different from what we believed; but it is always better to know the truth. Science does not have a monopoly on the truth nor does religion, as there are so many of them. But there are religions that are closer than others to the truth; and science, when looked at from outside the box of current trends of thought, does appear to support some religious truths and vice versa. Truth is more likely to do that, as it would prove the whole to be more internally consistent if so-called 'Revealed Scriptures' were to actually support a physical model and that a proper physical model would support the possibility of 'Revealed Scriptures'. Some may call this circular; yes, but it is a circle that, if demonstrated, requires science to accede to the veracity of those scriptures that prove to be in support of scientific truths and shows the two to be complementary. The model below delivers both, as it supports the possibility of revelation. This will be difficult for many to accept; some believe scriptures to be forgeries or the work of men and some 'faiths' have no 'Revealed Scriptures' but have documents that are authoritative but are acknowledged to be the work of great men - philosophers and men of great faith.

What is truth? The first time I heard that phrase was because of Sir Tim Rice's lyric in *Jesus Christ Superstar*, sung of course by the character of Pontius Pilate, "But what is truth? Is truth unchanging law? We both have truths; are mine the same as yours?" I found those very powerful words and they will always echo in my mind - thank you, Sir Tim. I believe that the search for truth must begin with ontology and that we must, first, discover what it is that exists before we can know, in an epistemological way, anything else. Why? Because ontology precedes knowledge, as I said above. So what exists? Only what we see? We already know better than that. Does God exist and, if so, can science support it? I believe yes and that which is below is my longer profession of that committed belief.

Above all, remember that I want to impart truth and believe that you, the individual reader and the public at large, deserve to know the truth. Only then can you begin to know your true place in the universe and better understand some of the more tragic elements of human existence. You have the right to know the truth and, in today's world, there are no censors to prevent this book from reaching a wide audience, although, I suspect that there might be areas where this book might be banned purely on some of the more political statements contained in the *Practical and Social Applications* section. What is below is what I believe to be the truth about how our universe works and why. I think you're prepared now for the adventure. I hope you understand and enjoy the ride; for it is a ride on the road to reality.

25

What is God?

One thing that most believers accept about God is that God is above and beyond gender. Although many languages, including English, commonly employ a capitalised, masculine, personal pronoun, "Him" or the possessive, "His", to refer to God, this is just a convention of past usage. That usage was derived from the language of the underlying source documents. In the case of The Jewish Scriptures and The Qur'an, the source is Semitic and in the case of The Christian New Testament, all accepted sources are Greek, although there may have been Aramaic originals of some, we have found none as of yet. Nonetheless, in these more egalitarian times, I believe that it is time to drop this masculinity from our reference to deity, if for no other reason than that we know better. Handily, the English language, because it is a Germanic language, affords us a gender-neutral option. Therefore, throughout this book, I will use the capitalised, English, neuter-gendered pronouns "It" and "Its" in place of the more familiar masculine usages; note that I maintain the capitalisation that has heretofore been granted to a personal pronoun that refers to God - I share the belief that the emphasis imparted by the capitalisation is, if nothing else, absolutely due out of respect. I don't do this, the removal of masculinity, as a slight to God - as if a human could emasculate God - or as a slight to any previous authors who maintained that usage but, rather, to reinforce that which is most commonly believed, that God is above and beyond gender and, to be fair, God ought to be referred to by genderless pronouns if pronouns are required and the language permits. In fact, I believe that God would approve of this, as some of the revelations given to prophets by God state clearly that God is without gender or is above that concept. Also, logically, as God depicts Itself as a one-of-a-kind entity, there is little point to Its having gender Itself; rather, It fully comprehends the concept OF gender and employs it throughout Its creation where necessary; but, It has no use for it Itself with respect TO Itself because It IS a one-of-a-kind entity.

One of the greatest problems regarding deity is that of definition. What do we mean by the term "God"? Western philosophers have debated many topics but none more pivotal than the question of the nature of God, i.e., that which makes an entity godlike. Equally, many atheists take great enjoyment from the confusion regarding a firm and reliable definition of what exactly is meant by the term "God", as it enables them to attack the believer in true Socratic fashion by questioning whether or not the believer has any idea of the topic of his belief and, then, use that as a means of vilifying all believers. This is, of course, not true of all atheists, but I have seen the tactic employed in countless arguments with atheists to know that it is a common ploy. Also, it does a believer no good to have no firm and reliable definition of God anyway, as, technically, it does eat away at the very foundation of their belief and may serve as a platform for doubt.

Saint Anselm, the 11th century priest and philosopher, was one of the first to try to use a formal, logical approach to determine the nature of God. His conclusion was that God was "that than which nothing greater can be conceived". To support that argument, he demonstrated that objects that are both imaginable AND real are greater than those that are only imaginable and, if God is "greater" than any (and every) other thing, It must be imaginable - as all would agree - AND real. Anselm's conclusive argument for the reality and greatness of God is fatally flawed, though. His critics pointed out that his argument naturally involves the limitations of the individual doing the conceiving because, if the conceiver can imagine nothing greater than a vast paradisiacal garden, does that mean that God is a vast paradisiacal garden that MUST exist somewhere? Quite clearly not. But Anselm was on the right track.

28

I have taken the liberty to correct that flaw and re-present the argument in plainer terms that are less liable to attack and/or debate. God is NOT "that than which nothing greater can be conceived" but God IS "that than which nothing greater can be". If you remove the conceiver from the argument, the statement is, in every way, perfectly correct. For now, the argument only involves existence and greatness; there is no third party involved and the subject of the statement is entirely about God. Because the statement involves only existence and greatness, one defining quality of existence, it is a more perfect ontological statement, as well.

The classical definition of God that most people would agree on, if they accept - even provisionally - the proposition of God's existence in the first place, is that God is omnipotent, omniscient and omnipresent. Discovering the real definition of God is vital to the premiss. The presence of these three qualities seems to be the classical litmus test for divinity in that, if an entity or object could be demonstrated to both exist in reality and have those three qualities, it would be nigh on impossible to refute that such an entity or object were God. But what do these qualities mean? Again, we have the problem of definition. We need to ensure that our understanding of these qualities is correct in order to know whether or not we have actually found them in any specific entity or object. I put forward the following definitions because they are sensible and logical and, I believe - expressly because of their sensibility and logic - they are more likely to be discovered in reality than other definitions I've read or understood.

Omnipresence is, perhaps, the least problematic of the three, although it can, and often is, viewed from more than one angle. Literally, "omni-" as a prefix, means "all" or "all-encompassing" and presence means "to exist at a particular place". Put together, it literally means "to exist everywhere". But it is deeper than that, now, because time is also a factor. Since Einstein's discovery of Special Relativity, we now understand that

29

space is indelibly joined to time in a 4-dimensional whole and, our understanding of 'presence' must allow for that new understanding of the link between space and time. For God to be omnipresent, It would need to exist in every point of space throughout all time. In more scientific terms, It needs to exist, at least, throughout the space-time continuum. I say 'at least', because there are, potentially, other dimensions - spatial, temporal or spatio-temporal - that may exist that are outside this space-time continuum. God would have to exist throughout all dimensions by this definition of omnipresence and it is this understanding that would cover the various String Theories that employ many more spatial dimensions than the Standard Model does.

Over the years, theologians have muddied the waters a bit and arrived at two distinct views on omnipresence: transcendence and immanence. A transcendent God exists not only within the Creation but also "above and beyond" it. In other words, Creation is a realm over which God has dominion and It can enter it at any place and at any time but, naturally, God exists in a divine, spiritual realm, Heaven, which exists as a parallel world that physical things cannot penetrate. On the other hand, an immanent God exists throughout the universe by placing a small part of Itself, sometimes referred to as a 'divine spark', into every thing that is in the universe.

It didn't take terribly long for someone to realise that a God that was both immanent AND transcendent was greater than a God that was only either one or another. Using Anselm's logic of applying absolute greatness to God, MY definition of omnipresence would also have to include both immanence and transcendence as the two required qualities to demonstrate omnipresence in a deity. Our scientific understanding of the universe would, then, tend towards more readily understanding an immanent God, that can have a portion of Its existence at every spatio-temporal point throughout the space-time continuum than it would towards a transcendent God, which, by definition, would have to have some of Its existence outside our known universe and, as we have no conclusive evidence for there being anything outside this 4-dimensional universe, Science will struggle against such a conclusion. To clarify, then, my definition of omnipresence, an omnipresent entity or object will exist both immanently - thus, at every point throughout the 4-dimensional, space-time continuum - additionally, it would have to exist transcendentally that is, if there are regions (or dimensions) outside or beyond or simply 'other than' our space-time continuum, then God must exist throughout those areas as well. There can be no place or time that could be outside the influence or, indeed, outside the very presence, of an omnipresent God.

Omnipotence is the quality of being all-powerful. But what does that really entail? Does it mean that God can make a human amoeba or a triangle with eleven sides? There are some that believe that to be the case. Some people feel that omnipotence must mean that God can do the impossible or those things that are logically self-conflicting. I say that's a ridiculous proposition and this is my reason. If God is real and can do the impossible and illogical then we can never rely on scientific findings because God could, at any given time, change the results of any given reaction. You would never even be able to say, unequivocally, that $2 = 2$ much less than $2 + 2 = 4$. If a real God can do that which is impossible or illogical then there can be no possible meaning to anything because it could be whimsically altered at any time and any place.

As the French dramatist Jean Anouilh pointed out in his play Antigone, "Nobody has a more sacred obligation to obey the law than those who make the law." God created the laws of physics in order to act in accordance with them. I believe that we can rely on scientific findings because our experiences have never been otherwise. That does not mean that we currently have scientific explanations for every event that anyone witnesses but it does imply that there IS an explanation for everything;

31

explanations, though, may not be scientific, as science relies on repeatable, empirical tests and, whilst there may be a perfectly reasonable and rational cause for an historical event, we cannot go back and change the events to test our theory. But even the fact that there is an explanation for everything comes with the caveat that we, as humans, may still never fully understand everything and, thus, may never discover the reasons for certain experiences or events. And I include in that vast area of events that, as yet, do not yet have explanations, but, potentially could, such things as: ghosts, extra-terrestrial life-forms, consciousness, telepathy, telekinesis and divine revelation. Extra-terrestrial life-forms either do or do not exist. Consciousness is impossible to refute, yet, so far, it eludes complete explanation. Ghosts, telepathy, telekinesis and divine revelation may have explanations that science can't explain, as they may be the kind of events that are unrepeatable and, therefore, like historical truths, have explanations, but not of the variety that can be tested by repeating a particular procedure. And the reason for THAT may be as simple as knowing that, in a 'repeat' scenario, the experimenter is NOT performing the experiment at the same time, and, therefore, different results may occur.

To be all-powerful doesn't mean being able to do the impossible but, by using Anselmic logic, it means not only being able to do all that is possible it also means actually doing all that is possible because a God who can do everything but chooses not to is less than a God who not only can do everything but also DOES it. So, to my way of thinking, in order for God to be omnipotent, It must be the sole individual directly responsible for everything that occurs throughout space-time and, of course, any areas outside space-time should they exist. God must not only be the Augustinian "first cause" but must be the one and only cause for any effect at any place at any time throughout the space-time continuum and any other areas should they exist. God must be the 'doer' of all that is done, for, if there is another doer, then, logically, God is not all-powerful because there is something ELSE that has done something. It is by having the quality or attribute of Omnipotence that forces God - yes, God is forced - be unique and is the quality that, above all else, signifies that God is singular and that there can only be one true God.

This will, no doubt, be abhorrent to those who realise this places God as directly responsible for every evil act. But that is an uneducated approach, certainly from a Platonistic, philosophical view. Evil is a perception; like all perceptions, it comes from within US and is projected towards an act we find in some way reprehensible. We perceive evil but that is not a proof of the existence of evil it is only a proof of our perception and humans are profoundly predisposed to misperceive. Humans tend to perceive evil in acts that are non-life-affirming and/or restrictive of civil liberty or events that are destructive of life and/or restricting civil liberty as discussed in greater detail in the more philosophical chapter *What about Sin and Damnation*. This division between acts that are considered evil because of their life-affirmation implications and those acts considered evil because of their civil liberty implications is very important because those of the latter variety are much more subject to cultural variance amongst our kind. In fact, offensive behaviour varies so radically within and between species that it should be blatantly obvious that evil, as well as good, is in the eye of the beholder. For example, if you smile at a cat that you don't know, it may run away because it saw you bare your teeth and perceived a threat. A 'thumbs up' sign in most places is OK; but don't use it in Brazil where it means 'up your arse'. Plainly, misconceptions based on perception are prevalent.

Also, we humans do not have, as God does, the third divine quality of omniscience, which means having universal knowledge; for, if we had, we would be able to KNOW good and evil rather than only hope to correctly perceive it. This knowledge of good and evil is the reason for God's trepidation regarding mankind's gaining it by eating of the fruit of the Tree of Knowledge between

Good and Evil in the Garden of Eden allegory. To be omniscient IS to be as close to God-like as is, perhaps, humanly possible (omnipresence and omnipotence are plainly not attainable in a human form due to our physical limitations); yet, I hold that omniscience is, most likely, unattainable for humans primarily due to their lack of omnipresence. For God, omniscience is made slightly easier to attain by virtue of omnipresence, as it is surely easier to know about everything if you are everywhere. But, for us poor humans, if we can come to understand that events occur in this universe, the reasons for which are known only to the whole because only the whole has access to its entirety throughout time, then we have come to terms with God with respect to good and evil and can place a firm trust in the knowledge that it is how we REACT to our perceptions of good and evil that is more important than any perceived good or evil.

Truthfully, omniscience, that is, "knowing everything", isn't enough; one must not only know the 'data' but understand it as well. Complete understanding we could call 'omniprehension'. It is by having omniprehension - knowing the meta-data that makes sense of the raw data - that allows God to act wisely and not just act wisely but with omnisapience, complete wisdom.

There is the computer-based analogy of a relational database where omniscience is: knowing the data within the rows and columns and tables of a database but not knowing the table names or the column names. Knowing the 'meta-data' is: knowing the table names and the column names and which columns are on which tables and the entire relationship of tables and columns throughout the database. Knowing the meta-data, then, is the root of all understanding – omniprehension - and understanding, for a computer, is simply knowing the entire database structure and knowing the relationships between all the tables and columns. So, in the case of God, God must know all the relationships between all quanta of energy throughout the continuum.

For simplicity's sake, I will include the concepts of both omniprehension and omnisapience in the term omniscience. Yet I needed to address those issues, as some may have thought that I overlooked those, more important, concepts. I have not; rather, I have employed a strict logic to the ideas I present specifically because I do not want to mislead the public regarding such vital information.

Philosophically then, God is an entity that is both transcendentally and immanently omnipresent, God is an entity not only capable of doing everything possible but actually performs every act and, as such, is omnipotent and God must know everything that could possibly be known, have complete understanding of that knowledge and be able to act with perfect wisdom, as It is omniscient. Furthermore, any entity that would fit the description above would also be that than which nothing greater could be and ties in perfectly with my modified Anselmic logic. So, now we know how to recognise a God when we see one. The question that remains is: Is there an entity that phenomenologically presents a case that fits this philosophical ontology? I purport that there certainly may be and, more likely, probably   is.

Who is God?

Theistic beliefs come in two main varieties: monotheism and polytheism. And truly it isn't difficult to see how, if you believe in 'the concept of God', you will either accept that there are many of them or that there is only one. Any system that maintains one supreme god is monotheistic and any system that maintains more than one is polytheistic. But most polytheistic faiths also have a concept of Chief God. Zeus of the Greeks, Jupiter of the Romans and Odin of the Norse all fit this concept, as they were unbeatable. Dvaita (dualist) Hinduism, which purports 3 main gods - Brahma, Shiva and Vishnu - and thousands of lesser gods, has, at its Advaita (non-dual) core, the single, unique Brahman and the concept that there is nothing 'other' than Brahman. The old Egyptian pharaonic religion had Amon, a God that existed before any others, self-begotten. And, let us not forget Guichi Manitou (Great Spirit) of North America. So, even polytheistic religions tend to hold one god as greater than others. It seems, then, that a belief in one god, supreme above all, is an almost universal belief among those of a theistic paradigm. Logically, you cannot get past the concept of supremacy (omnipresence, omnipotence and omniscience), which can only be held by a single entity.

Atheists either maintain 'no belief in any god' and there are also those who will ask a believer for their definition of God and then attack that in true Socratic form and those that specifically believe that there is no god, as it is unnecessary because science can (and will ultimately) offer rational explanations for all phenomena. But that leaves a hole regarding those phenomena for which science, as yet, offers no explanation, for example, the answer to the questions, "Where was the energy before the Big Bang?" or "What are ghosts?" Science takes a view that the

37

observer is deluded, in the latter case, often with no investigation whatsoever; and, in the former, simply has no explanation beyond "It just, sort of, happened." Today's scientific view, although it is never stated openly, holds that no 'scientific' result can ever imply deity. Science, in accordance with that unspoken dictat, will simply not let God enter the equation. And, because of that, shuns as delusional those events it cannot explain or refuses to investigate. This is, in a strict sense, very bad science. These events may simply be extremely rare events or are, in some way, unrepeatable. Atheists hold that there is no conclusive evidence of God and religionists hold that there is, equally, no conclusive evidence to disprove God's existence. And, so, the battle lines are drawn. Both sides committed to their opinion supported by an overwhelming lack of evidence. The agnostic logicians look on and are completely bewildered by such profoundly held positions based on 'a lack of evidence', when they know, beyond *any* doubt, that no logical conclusion should be drawn from a lack of evidence.

One of science's greatest mysteries is quantum entanglement. The concept that, when one measures the spin of one of a pair of particles that had a single origin, that the spin of the other particle is affected immediately, irrespective of the distance between the paired particles. There is, supposedly, a demonstrable 'immediate', i.e., faster than light speed, effect. Einstein didn't like this and referred to it as 'spooky action at a distance'. He was convinced that the speed of light was the limit and that nothing could exceed that. But, is there a way that could explain both, that is, a physics where certain aspects of quantum dynamics demonstrate evidence of a hidden unity between seemingly separate particles yet no actual particle can move faster than light speed? Is the answer half staring us in the face with the phrase 'entanglement'?

String theory offers a very elegant solution to physics by granting that there are more than just our 4 dimensions of space-time. It purports that there are, at least 6, if not 7 in M-theory, additional dimensions to our universe. In fact, the original String Theory - the first purported and oldest of them all - maintained that there were 26 dimensions. Coming back to the more modern string theories, these additional dimensions are found in an extremely small, compactified, space - called the Calabi-Yau space - that exists throughout our 4-D space-time on the order of the Plank-scale. String theory states that energy is in the form of tiny, Planck-scale sized strings that, when they vibrate at different rates, produce the 'effects' of the various sub-atomic particles. The Planck-scale is extremely small; the size of a string of energy, in string theory, is to a proton about what a proton is to our solar system. But, string theory, as of yet, says nothing about what energy might do in these other dimensions.

It is reasonable to assume that these dimensions are not left unused and that energy acts as much 'there' as it does 'here' in our 4-D space-time. Considering the long-assumed axiom that energy is conserved, one would think that energy is conserved in all dimensions and not just the 4 more obvious ones. And, that, in all likelihood, energy exists, simultaneously across all 10-dimensions. That is, that these strings of energy are 10-dimensional and that what we observe in our 4-D universe is only that part of the overall string that is referenced in these 4 spatio-temporal dimensions. If this is the case, then that would result in an overall conservation of energy throughout the entire multi-dimensional universe. One of the hopes of string theorists today is that some of the experiments performed at the Large Hadron Collider will demonstrate a hitherto unobserved phenomenon - that energy would either appear or disappear.

As science is quite happy with the concept that energy is neither created nor destroyed and that, rather, it is conserved, such a gain or loss of energy would destroy that belief given the Standard Model of physics; however, to a string theorist, it would offer immense hope. The reason is that a gain or loss of energy could be explained as energy that has moved from the Calabi-Yau space into this 4-D universe (a seeming 'gain' of energy) or that energy has escaped this 4-D universe and has moved into the Calabi-Yau space (a seeming 'loss' of energy). An observed loss or gain of energy, then, is evidence towards string theory as being the better explanation of reality and that what has truly been observed in such a gain or loss is that energy has been conserved within the larger, multi-dimensional universe and that the Standard Model, which does not predict this kind of gain or loss, would have another great hole ripped into it. Since string theory has more 'places' that energy can be, it can more easily afford such an observed gain or loss than either the Standard Model or Quantum Mechanics and this kind of behaviour, if observed, would go a long way towards explaining the so-called 'quantum flux' or 'vacuum energy'.

Another of the problems, though, with String Theory is that, whilst the general shape of the Calabi-Yau space is known (in fact, it's a 'Calabi-Yau' shape!), the exact length of each of these other dimensions is unknown. What allows the model to work on paper is that the entire Calabi-Yau space needs to be of a certain size; in truth, its overall size only needs to be within a certain range of sizes for the mathematics to support the universe as we observe it. But, getting into that next level of detail - the lengths of each dimension - is, again, an unknown; but the system will work as long as each dimension is within a certain range of lengths such that the overall size of the Calabi-Yau space is within the required range.

To date, as far as I can tell, no one has tried to determine the length of even one of the Calabi-Yau dimensions. This is due, primarily, to the fact that they are so small as to be impossible to observe. Secondly, it is extremely difficult to tell what advantages are to be had by having one dimension of X length and another of X + 1 Planck-length as opposed to the other way around - and that's just dealing with 2 dimensions out of 6. Trying to discover which conformation across the thousands of possible conformations is the 'best' is, at present, considered by even string theorists to be

40

an almost impossible task. However, I believe they have done what so many scientists before them have done and have completely overlooked some of the evidence that rests before their very eyes.

The concept of Quantum Entanglement is, I believe, a very revealing piece of evidence. Besides the fact that it is a phenomenon that can be repeatedly demonstrated in a laboratory given the right conditions, I believe it sheds light on the conformation OF the Calabi-Yau space. If, for example, just one of those Calabi-Yau dimensions were of exactly the Planck length in length - the absolute smallest length possible given our known laws of physics and the effective size of a 'point' in our universe - then that would force all the strings that exist to join at that point in the Calabi-Yau space. Looked at in a coordinate system, that one Calabi-Yau dimension offers only one coordinate, thus all strings would share that coordinate in that dimension. And as it is in the Calabi-Yau space, it is in a space outside of our line of sight and, as it is also outside of space-time, thus outside of time, it is, therefore, an eternal or atemporal join. Given a universe of stringy energy in which the strings are joined in the Calabi-Yau space, you would EXPECT to find quantum entanglement, as the quanta are, in fact, just 4-D extensions of a larger, unified, 10-dimensional object of stringy energy. For, if strings must exist in all dimensions (due to conservation of energy expressed across a 10-dimensional universe) and, in one of those dimensions they are joined (by virtue of one dimension affording only one, effective coordinate 'point' of reference IN that dimension), there is only one, actual, object that exists: a 10-dimensional object of stringy energy. It seems to me that this conformation of the Calabi-Yau space solves Quantum Entanglement because it results in describing the entire universe as one single object and, whenever you affect one part of that object, you affect the rest of it because every action has an equal and opposite reaction.

With just this simple tweak, the Calabi-Yau affords us a rectification between the Standard Model and Quantum Mechanics with respect to Quantum Entanglement. It also defines the length of one of the Calabi-Yau dimensions, which SHOULD make the task of defining the lengths of the others easier, as it's a matter of 1 down, 5 to go. But, after allowing for one dimension to be exactly the Planck-length in length it has the knock-on effect of turning the entire universe into one object. And, if that is a true reflection of reality, what other realisations can be made regarding a universe so described? Are there any philosophical implications that can be derived from that configuration? I believe there are vastly important implications from this configuration; therefore, I am compelled to state them.

Firstly, this single object of stringy energy is everywhere energy is throughout all dimensions of space, the Calabi-Yau space, which is outside of time and our 4-D universe that contains time; so it fits the description of what is required for both transcendental omnipresence and immanent omnipresence. Secondly, if there is only one object that truly exists in the universe and all other 4-D objects that appear separate only appear separate because of the geometry of the universe itself, then, with only one actor in the system, that actor is responsible for every action that takes place throughout all dimensions, therefore, it can be reckoned as omnipotent, as there is no other actor that exists in the 'Bigger-picture' reality. Thirdly, irrespective of how consciousness works, with only one actor in the system, all consciousness, in some way, must be the consciousness of the one thing that exists; therefore, it fulfils the requirements for omniscience, as there are no 'other' knowers in the universe save the single entity and the appearance of disparate conscious beings in our 4-D universe becomes an effect of the way energy is bent and twisted around the several dimensions that exist. Therefore, it is fairly deduced that a single 10-dimensional object of stringy energy (For the moment, I'm sticking with the 10-dimensional model, but irrespective of the number of dimensions, if one Calabi-Yau dimension is exactly the Planck-length in length, then the resulting configuration leads to the same conclusions!!) joined by virtue of exploiting a simple geometric configuration, results in defining the universe as omnipresent, omnipotent and omniscient and, thus, qualifies it to be called God. And, finally, God falls out of the equations. Science runs open-mouthed in absolute distress unable to express its grief and horror.

This was the epiphany I stumbled across several years ago when I was trying to use String Theory to rectify the problem of Quantum Entanglement (spooky action at a distance) and the Standard Model. Once I'd made the discovery that it was possible and even plausible (because it solved the problem) to solve the problem by shortening the length of just one dimension of the Calabi-Yau space, the philosophical implications hit me very quickly and very hard. As I was never one to accept the unwritten dictum that God cannot be a scientific result - predominantly because I know the dictum itself to be unscientific - I was still astounded by the discovery. In fact, I was truly awe-struck. Had I just seen, for the first time, a situation where an individual was only looking for a scientific explanation, yet God simply fell out of the equations? I believe I had.

The model I suggest above solves for Quantum Entanglement and suggests that C, the speed of light in a vacuum, remains as the limit of speed in our 4-D universe; however, it also suggests that certain aspects of information regarding energy in this 4-D universe is relayed via the connection in the Calabi-Yau space between quanta, in this case, spin or angular momentum. It also suggests that Entanglement is a natural state of affairs for ALL quanta with respect to any other quantum but that it is only observed under conditions where the individual relevant quanta have been isolated and the environment controlled under laboratory conditions; in other words, the only time we can observe Entanglement is when we control the conditions and protect the quanta from interference by other quanta - excepting, of course, those we use to measure and, thus, cause the reaction to take place.

So, Einstein was right to balk at 'spooky action at a distance' insofar as it pertained to the Standard Model; however, String Theory can, as shown above, solve the problem and make Quantum Entanglement a normal, natural and expected observation of a universe that, in reality, is supported by a String Theory paradigm. The 'new' problem for scientists, now, is to understand that, by using the model that explains Quantum Entanglement as being perfectly expected, it also asserts that the universe is, in fact, God. A God that is one, single entity, just as It is described in countless revelations that science refutes as simply 'tales of the ancients'.

But is this *exactly* the kind of deity that is purported by those books that claim to be direct revelation? Are there more philosophical implications that can be drawn from this model that match up with other statements made in these revelations? Or, are there statements made in these revelations that are made impossible by this model or does this model specifically refute statements made in revelations? Firstly, can the model even support the concept of direct revelation? If it cannot support the concept of revelation as being possible, then, therein lays a huge problem. Does this new model shed any new light on consciousness? It can quickly be seen that, even though the model I propose solves one question, like so much of other science, it raises far more newer questions and the investigation to discover or uncover the answers to those questions becomes the next chapter of the book.

Which String Theory is the Best?

Before we answer some of those more interesting questions, it's worth spending some time on discovering which of the various string theories available tends to be the most likely. As I'd mentioned above, there are three main theories that differ in the number of dimensions they propose. There are four different theories that propose 10 dimensions, one that proposes 11 dimensions and the earliest and most mathematically complex theory that purports 26 dimensions.

Most modern string theorists have practically abandoned that original 26-dimensional theory because it supported and predicted the existence of tachyons, particles that move at faster-than-the-speed-of-light and, naturally, this is in direct conflict with the Standard Model and in direct conflict with observations. That is, to date, no one has observed a tachyon. However, I don't believe that not observing a tachyon at this point in the evolution of our 4-dimensional universe is proof or even 'evidence against' their existence. It is extremely plausible that tachyons, as fast moving as they are, could well have existed in the past and have, since then, changed into something else by now. Modern scientific belief is that tachyons, under certain conditions, actually condense into Higgs bosons, the proposed and as yet unseen 'particle' that carries the force of gravity.

I'm now going to enter a realm that may, indeed, turn many scientists or science-minded readers a bright shade of red and they may throw this book across the room in a rage of defiance against what I present as evidence. Be that as it may; damn the critical torpedoes of institutional science and/or religion - I will not let them deter me from my course. I have already stated that scientific 'truths' are not the only form of truth that there is and I'm completely confident that THAT is, indeed, the case. I gave, as an example, the case of historical truths where we know we have causes for events but there is no way to go back and test our theories to know, for sure, that the causes we believe are the true causes of events, are, in fact, the actual cause(s). Science demands the quality of repeating an experiment and getting the same result; however, life doesn't always present us with the possibility of exactly repeating any given scenario and I supported that by stating that, if anything, it is obvious that, in any secondary experiment, time has moved on and that alone causes a new, temporal environment.

Now, here's the point at which many scientifically-minded people as well as staunch atheists may throw this book across the room, although I implore them to not do so and to continue reading in order to be fair to the entirety of the material I present. I purport that scripture, especially 'revealed scripture' must be considered as a form of evidence. It is NOT scientific evidence, let me make that clear; but, it is evidence just the same. If I give testimony in a court, my evidence is not necessarily scientific evidence but, simply, my bearing witness to my own observations. In the case of 'revealed scripture', we have a situation where what is proposed is that God Itself is giving witness to Its own observations. And, if that premiss is accepted, then, we'd be foolish to ignore scriptural evidence. Besides, if one is trying to demonstrate what happens when science meets religion, then one must not only look for hints of religious truths in science but, also, look for hints of scientific truths in religion; truly, it's only fair.

In this case, It is Genesis 1 to which I wish to refer. It describes a 'separation of the waters' such that the 'upper waters' became Heaven and the 'lower waters' became Earth. We must also understand that many of our modern concepts have no equal in earlier languages; this is a fact. There was no Ancient Hebrew word for 'telephone' or 'microwave oven' because neither these concepts nor their material counterparts existed in those times.

This is no fault of the language used; it is just a result of when technologies or certain aspects of knowledge entered human experience.

Modern science has no answer as to what exactly happened to the anti-matter in the early universe. Rather, it purports that, a certain amount of time after the Big Bang (in fact, only the briefest fraction of a second), the anti-matter 'went away' and what was left was primarily what we call matter. What we do know, though, is that, while both were in existence, they acted in accordance with the laws of fluid dynamics, which left them in a state where they were continually colliding with one another and annihilating each other leaving a 'soup' that was predominately 'light', i.e. varying wavelengths of electromagnetic radiation. As soon as any condensation occurred and a 'particle' was formed, it would find its way to a similar anti-particle and the two would annihilate one another until that mysterious parting of the ways between the matter and the anti-matter.

I suggest that 'water' was the closest term to 'fluid' that ancient Hebrew had and, as it had no words for matter or anti-matter, the scripture referred to them by calling them two different types of 'water' - upper and lower - because they acted in accordance with fluid dynamics. Furthermore, I purport that, in God's great wisdom, It separated these two varieties of energy so that one of them - in this case, that which we call matter - was left in a state so that it could condense into higher particles in relative peace without the constant threat of continual annihilation leaving the evolution of this universe to begin to unfold. But what did God do with that anti-matter? Referring back to the same line of scripture, the anti-matter was the 'upper waters', which became Heaven and the matter, the 'lower waters', which became 'Earth'. Now, in this usage of the term 'Earth', again we run across a limitation in the ancient Hebrew language. I believe that, in this particular usage, the term 'Earth' does not refer to the planet we live on but, rather, it refers to the entirety of the 4-dimensional aspect of our universe that we refer to as the space-time continuum.

That 4-dimensional space-time continuum expands. We've known that since Edwin Hubble discovered it in 1929. But what does it expand through? It must expand through some medium and I believe it is that medium that is responsible for gravity. Why do I say this? Because I believe that the anti-matter was taken from the mixture that first existed and was set up as a barrier far away from where the matter was. This barrier is negatively curved, that is, it is shaped like a doughnut (torus) with a tiny, Planck-length sized hole in the middle. Just after the Big Bang occurred, the tachyons sped out with increasing speed and something was required to contain them in order to conserve their energy. So God took the anti-matter (separating the waters) and set it up as a toroidal boundary for them to run into. When they did, they condensed into what tachyons condense into when they meet a boundary through which they cannot pass: Higgs bosons, the particles responsible for imparting the gravitational force. In other words, the tachyons - spreading out in all directions from the Big Bang - laid out a huge network that eventually (once they had condensed) created one, very large Higgs particle structure within the 'space' between the matter and the anti-matter boundary. It is that Higgs-structural medium through which our 4-D space-time expands. In other words, this 4-D continuum is, in a sense, inside the Higgs boson, which is why gravity works throughout the universe and why we cannot find a Higgs particle within the 4-D continuum.

For the Higgs boson to impart its force equally throughout our 4-D space-time, it could well exist in a region that is perpendicular to those 4-D directions, thus the Higgs 'field' or structure could - and I believe DOES - exist in the dimension through which our 4-D space-time is expanding. The reason that space-time expands rather than contracts is because the barrier towards which it expands is anti-matter. The opposite charge of the anti-matter causes a constant electromagnetic attraction on the matter and drags that matter from the past into the future explaining the reason why space-time expands rather than contracts, which has been a very perplexing question for physicists for years and a question that this model answers. And, as far as I know, this is the only model of physics that actually proposes an answer to this question that is both reasonable and plausible.

It is on the very edge of that expansion where we exist. On that very bow wave of expansion is the only place where 'motion' takes place and the only place where 'motion' can be experienced. The expansion of space-time IS motion and all motion is because of this expansion. Anything that is behind that bow wave of expansion is in the past and anything that is in front of it is in the future. The main reason that we cannot see into the future is because we experience our existence right on that edge of the expansion of space-time. For, as soon as we move, we enter the future in a new 'present' and, just as quickly, the fleeting moment gets forever locked into the past. It is memory alone that allows us to peer into the past by remembering it and memory is one aspect of consciousness. Rocks do not remember the past anymore than they can see into the future. It is the attribute of consciousness in humans (and, perhaps other species) that allows us that glimpse into the past and to be able to assemble or integrate the individual slices of 4-D events into a flowing whole. I will speak more about this aspect of consciousness later, as it gives us some vital evidence as to how consciousness works and also lends further credence to the string theory paradigm as a whole.

Because the original 26-dimensional string theory is the only one that predicts the existence of tachyons, it is also the only one that allows such a description of the early universe as I provide above and it answers some of the more mysterious questions about the early universe. It answers the question: Where did the anti-matter go? It answers the question: Why was the anti-matter separated from the matter? It answers the question: Why does our universe expand rather than contract? It answers the question: What happened to the tachyons? It answers the question: Where is the Higgs boson? It answers the question: Why can't we find the Higgs boson? It gives evidence to our experiencing motion because we can only experience anything if our existence is noticed by our consciousness when we are on that bow wave of the expansion of that continuum. All these questions this model answers and neither the Standard model nor Quantum Mechanics affords us these answers. Therefore, this model, because it answers so many more questions, begins to look more and more likely. But, of course, we still need to delve further into the details and reveal even more; so let's do just that!

What Does Energy Look Like?

String Theory defines two different types of string-like energy: open strings and closed strings. Open strings look like lines and closed strings look like circles. The electro-magnetic, weak and strong atomic forces are, in String Theory, handled by open strings and gravity by a closed string. This is the commonly proposed theory. Open strings vibrating at varying rates account for the multitudinous variety of sub-particles that compose matter and anti-matter; however, the Higgs boson, the particle that supposedly carries the gravitational force is purported to be a closed string. In Figure 1, below, I show what these strings might look like if we could see them. You see, they are very simple; a line and a circle.

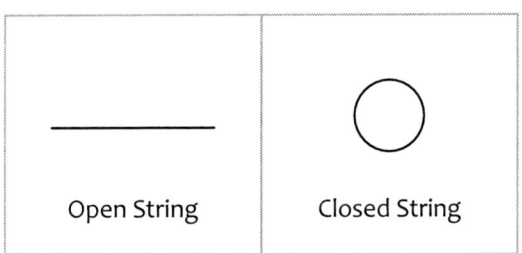

Figure 1: Open & Closed Strings without Vibrations

In Figure 2, below, I show what these strings might look like if they were in a state of vibration. It is the varying rate of the vibration of an open string that makes the string 'appear' to act as one of the various sub-particles of matter. Exactly how each sub-particle is accounted for, while possible to provide, would probably bore you;

the truth of the matter is that String Theory, in its present form, can account for the universe, as it has been proven mathematically; so, I accept it as a base premise from which to build.

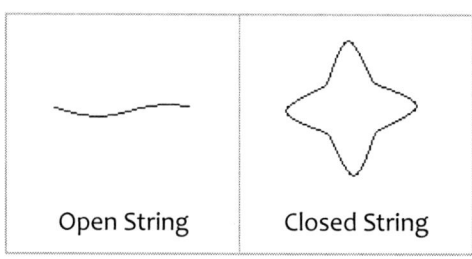

Fig. 2: Open & Closed Strings Vibrating

See how the vibrating strings are, topologically, still obviously open and closed, but they demonstrate a slightly different shape. It is that different shape - caused by the rate of vibration - that gives them their power to appear as different particles and allows them to interact with one another in many, countless ways.

But, the question remains: How do these open and closed strings appear when seen in 10 dimensions? What we see above only accounts for the 4-dimensional aspect - the aspect that we see in our space-time continuum. It does not depict, in any way, what energy does in the six dimensions of the Calabi-Yau space, the 'other' extra dimensions that String Theory affords. If energy is conserved, then it is conserved in all dimensions and it would be ignorant to assume that energy within the Calabi-Yau space does nothing when we can see the incredibly varied things energy can do in these 4 spatio-temporal dimensions. As the Calabi-Yau space is not directly connected to time, as we shall see, its structures are eternal; the dimensions of the Calabi-Yau space are purely spatial. And these facts, naturally, lead us to other, perhaps, more basic questions:

1) How do these 10 dimensions inter-relate with one another?
2) What do strings look like when viewed in 10 dimensions?
3) What does energy do when it is in the Calabi-Yau dimensions?

These are very difficult questions, yet the answers to them are pivotal if we are to understand how energy works through these several dimensions. One thing that modern science has yet to really address is the concept of consciousness. They are very pleased with showing us Functional MRIs (Magnetic Resonance Imaging) and telling us that we are seeing thoughts when we see movements of energy pass from one part of the brain to another; but I'm not convinced. Rather, I think what we are seeing in an FMRI is the physical aspect alone - the physical interface TO consciousness, rather than consciousness itself and here's why I think that. I believe that the consciousness that we have is 2-dimensional. It is a 2-dimensional slice of an overall 3-dimensional 'Greater Consciousness' that is God's. And these 2 dimensions of consciousness cannot be like the spatio-temporal dimensions, so they must be handled by the Calabi-Yau dimensions. This, of course, means that they are eternal and that implies 'consciousness after death' (if not also prior to incarnation!) because the structure is atemporal, i.e. timeless.

But why do I say that our consciousness is 2-dimensional? Simply because we need one dimension in order to grasp the moment - to apprehend the 4-dimensional 'now' - in a place OTHER than those physical dimensions, which are already filled with what we are attempting to apprehend; and, we need a second dimension that enables us to integrate those moments into a flowing whole such that we can perceive a flow to events. Without being able to integrate individual moments, we would be unable to perceive a flow to time or a flow to events and, since we do perceive these things, there must be a mechanism that allows it. To have 2 dimensions that are not of the usual spatial dimensions of our 4-D space-time but dimensions dedicated to consciousness is a simple mechanism to allow individual processing of 4-D events. These dimensions of consciousness, though, must be unlike the normal spatio-temporal dimensions. They are dimensions of the mind—although they are more like spatial dimension, in that they are atemporal (being in the Calabi-Yau space), they describe an 'area' of consciousness; they are still enough unlike the other types of dimensions that I feel safe in defining them as 'dimensions of consciousness'. In this regard, they are a third type of dimension. By connecting to those dimensions - or allowing an interplay between those dimensions and the physical dimensions - an interface is created that allows us to use our consciousness; and our brains and central nervous systems are structured in such a way as to set up this interface to what I call "Consciousness Space". What the FMRI sees is the physical aspect of that interface and not consciousness itself; consciousness itself, I purport, is contained in the Calabi-Yau space and 3 of the Calabi-Yau dimensions are dedicated to and support "Consciousness Space".

What, then, happens in the other 3 dimensions of the Calabi-Yau space? I conclude that it is these other 3 dimensions that contain the definitions for abstract concepts - Plato's 'Forms', if you will - and this becomes a fourth type of dimension that is unlike the other types. So, the four types of dimensions that I propose exist within the combined space-time and Calabi-Yau space are temporal, spatial, consciousness and abstract. The abstract dimensions are, like the dimensions of consciousness, more like spatial dimensions because they are also atemporal, but they are different enough, I believe, as to warrant being called 'abstract dimensions'. In a way that parallels Plato's belief, I believe abstract concepts have their definition and their very existence defined in 3 of the Calabi-Yau dimensions. Because the Calabi-Yau space is atemporal (that is, without time as a component), all things therein defined are eternal, so these abstract concepts and consciousness itself are defined and have their existence in a space that affords them eternal existence.

If I'm correct in this idea, then Plato was correct in thinking that the abstract archetypes are perfect and eternal and consciousness, too, is eternal, which, as I mentioned above, lends credence via implication to the concept that there is 'life after death'. More properly, it isn't life after death, but a continuation of consciousness after physical death. Death, like birth (or 'incarnation') becomes nothing more than a transitional state when consciousness loses its interface to the physical realm - the 4-dimensional space-time continuum (the opposite process of birth is when the interface to the physical realm is formed and that process is 'incarnation', although it might be more fair to say that the interface develops during gestation). The real home of consciousness is the eternal space where it actually exists, which is in the Calabi-Yau space.

Thinking, then, becomes a procedure not unlike fishing with a net. When we think, we cast a net of consciousness into the pool of abstracts and return the results of that casting back into the brain. If this is the case, then it also implies that there exists an interface between the "Consciousness Space" of the Calabi-Yau and the "Abstract Space" of the Calabi-Yau space just as there is an interface between consciousness and our physical bodies. The interface between consciousness space and the abstract space would be what we call 'the mind', the reticulum or net, which we cast into the pool of thoughts contained in the 'Abstract Space' when we think.

A difference in IQ could, then, be analogous to differences in the 'net' that we cast into the Abstract Space. For example, a person with a higher IQ may have a net with smaller holes and, therefore, when it is cast into the pool of ideas, it catches more, as fewer, smaller ideas are trapped by the net than by a net with larger holes. The net with larger holes allows fish to escape, whereas with smaller holes, fewer fish can escape. It is also possible that the net of an individual with a higher IQ is larger and covers a larger area than that of an individual with a lower IQ, thus catching more ideas. Either of these two concepts is reasonable and the truth could indeed be a mixture of the two.

Perhaps by now your eyes have glazed a bit from all this new information. I've discussed two different types of space, 'Consciousness Space' and 'Abstract Space' and these are not altogether new concepts; rather, they are a new way of referring to older, classical concepts. Abstract Space is, almost in every way, a modern String-Theory answer for where Plato's Forms exist. Consciousness Space is what allows the individual to relate the abstracts found in Abstract Space into this Physical space, the kind of space with which we're most familiar, in a way that describes our overall design as Descartes' 'thinking thing.' If I've moved to quickly, then please allow the schematic analogies below to draw pictures in your mind. What I've done is to accept the premiss of String Theory and stand on the shoulders of those who did all the busy work to prove it on paper. My job is to see how far I can see given their premiss; but, because I've already done that, I would like to show the world what can be seen from that vantage point. So, please, stay with me.

Now that I've described all these different types of spaces and alluded to how they all fit together into one, single united whole; I need to show you, don't I? I think it best to simplify matters and use a form of schematic to demonstrate what I mean. In the figures below, I will use a stick-and-ball style of schematic where the balls (or dots/circles) represent the dimensions themselves and the sticks (or lines between the dots/circles) will represent the way in which the dimensions relate to one another.

In this first diagram, figure 3, I represent the Dimensions of the Calabi-Yau space as two groups of three dimensions each. One group relates to the 3-D Abstract Data Space, where thoughts and ideas themselves are defined and the other is the 3-D Consciousness Space, where consciousness itself is defined and its existence is maintained. Notice that I have given names to each of the dimensions and this may help you understand how these dimensions are utilised.

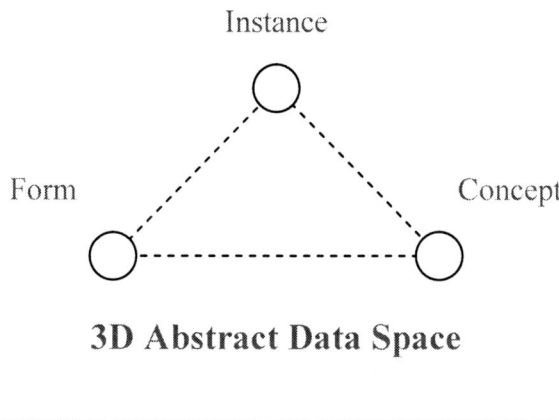

3D Abstract Data Space

3D Consciousness Space

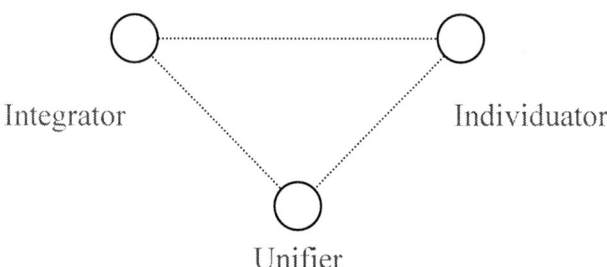

Figure 3: Abstract and Consciousness Spaces

In the Abstract Data Space, I have one dimension that is called 'Concept'; it is this dimension that defines the abstract in general terms. For example, we could have a definition for "container" here. The dimension named 'Form' is used to further refine the general concept into such things as "cup" or "bowl" or "Tupperware box" or as specific as "a crystal, fluted glass with ornate engravings". The dimension I've called 'Instance' is what defines where and/or how and, indeed, whether or not this object exists. A concept could be defined as only imaginable or able to exist in dreams but not in our 4-D space-time continuum. It could, though, also be defined as able to exist in our 4-D 'reality'. I propose that the topology of this 3-D area is what mathematicians refer to as a 'pointless topology'. That does not mean that it is an area without reason (as in 'it's pointless to continue discussing this'), rather, it means 'without points' in the mathematical sense of 'point' - topologically, it is perfectly smooth. In other words, its topology affords it the ability to retain and define an infinite number of abstract ideas and to define them to any level of detail. How this region is coded, though, is beyond this author, but I know that a proper, pointless topology can afford any 3 dimensional area the ability to define an infinite number of ideas and that is all that is necessary to understand and accept with regard to this region. To go into the details of this is not necessary, although if a reader is (or enough readers ARE) interested, I'm sure I could be contacted and a lecture organised.

With respect to the 3-D Consciousness Space, I have a dimension called the 'Individuator'; this is the dimension that is used by each of us to apprehend a given moment of space-time and it maintains our individuality with respect to our unique continuity of perception. It is this dimension, the Individuator, which grants us our individual train of thought and, truly, defines us AS an individual. The 'Integrator' is the dimension we use to integrate these individual moments apprehended by the Individuator into the flowing stream of consciousness that we perceive. Without the Integrator, we would not be able to access memory or know of anything other than the 'now' itself. The 'Unifier' is the dimension that is only one Planck-length in length and it is in this dimension that all the dimensions except time are joined together. This unifying dimension MUST exist in 'Consciousness Space' or the whole, which is God, would not be aware of its entirety and, as God is defined as omniscient, it is in Consciousness Space where everything must come together or, by reverse analogy, it is from this point that all springs forth. By placing the unifying dimension in Consciousness Space, self-awareness for the One is possible and at a quantum level, awareness is maintained throughout all types of dimensions by placing the unifying dimension within Consciousness itself. For now, please accept it; however, I will speak more in depth about it later. For the moment, I just want you to understand that this is where this unifying dimension is and, I'm sure that countless, clever readers have already foreseen many of the reasons why. Since we have alluded to the fact that the unifying dimension lies within the 3-D Consciousness Space, we may as well connect a few more lines to display that fact and that is what the figure below, figure 4, does.

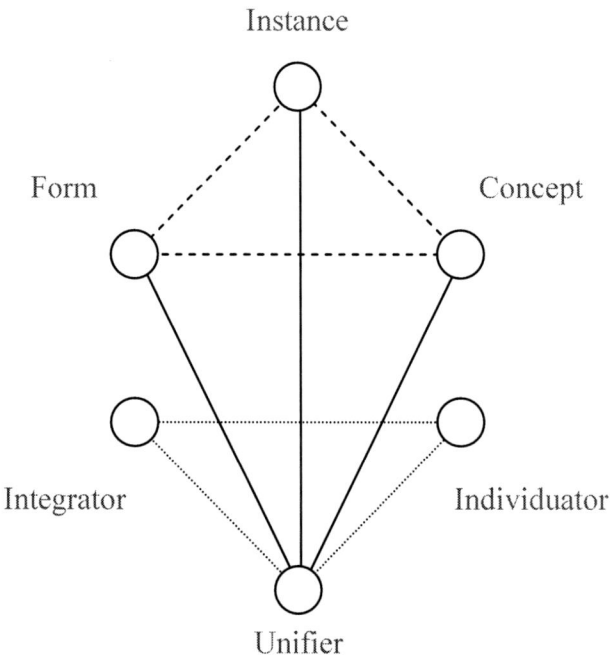

Figure 4: Abstract and Consciousness Space with Unifying Connections

I've used black lines to show the extra three lines that represent the connections between the 'Unifying Dimension' and the three dimensions of the Abstract Idea Space. Now, that unifying dimension has five lines that lead to each of the other five dimensions of the Calabi-Yau space. So, heretofore, we have outlined the schematics of the Calabi-Yau dimensions and defined what each one does; however, we haven't touched on that which is far more familiar: our 4D physical space-time continuum. In order to define the space-time continuum in a similar way, we need to have four balls and they each need to connect to each other.

Three of the balls would represent the height, width and depth of the spatial dimensions and the fourth ball would represent time, the temporal dimension. In order to reflect the reality of Einstein's Special Relativity, these four dimensions must each be joined to each of the others. Below, in figure 5, I attempt to show just that.

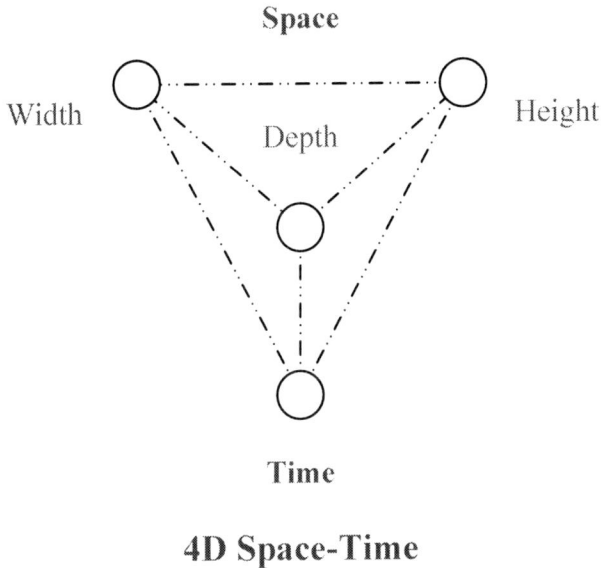

4D Space-Time

Figure 5: The Space-Time Continuum

In truth, it doesn't matter which of the three dots represent which of the three spatial dimensions in the figure above regarding height, width and depth. The important aspect is that all three are covered and contained within the diagram. I've chosen to place time, though, at the bottom, as I think that it is easier for the viewer to see how each of the three spatial dimensions join to it as well as to one another. But, now that we have a representation of our 4-D space-time, we should connect it to the Calabi-Yau dimensions in order to get a better appreciation for the whole. Also, I have to add in the extra links that the unifying dimension implies. In this case, the 'Unifying Dimension' that exists in Consciousness space must link to each of the spatial dimension of our 4-D space-time. It needn't link directly to time, though, as the links afforded by the three spatial dimension of the 4-D space-time to time is enough to bind time to the whole whilst still keeping it a safe distance from the atemporal dimensions that, otherwise, would be spoilt by a direct link to time. Below, then, in figure 6, is the conglomeration of figures 4 and 5 above with the extra lines that join the Unifying Dimension to the spatial dimensions of the 4-D space-time continuum added in black.

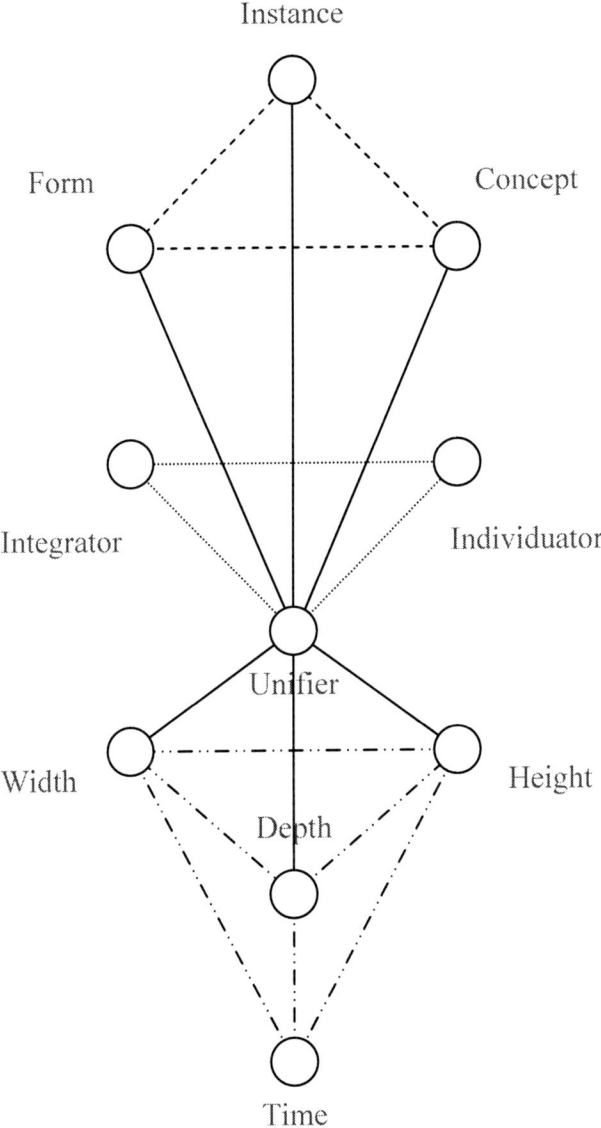

Figure 6: Figures 4 and 5 Joined

Now, the only things that are not represented in our stick-and-ball schematic are the two interfaces: the one between Consciousness Space and the Abstract Idea Space and the one between Consciousness Space and the physical, 4-D space-time continuum. If we add those two interfaces, say, with grey lines to make them more easily seen, the whole configuration is complete. In the drawing below, figure 7, I've changed all the unifying lines to black to indicate their connectedness, yet you should still understand that, although some were formerly dashed to indicate their inclusion in 'Consciousness Space' they still hold that role; I've only re-coloured them to show how they actually stem from the 'Unifying Dimension'. Also, I've taken the liberty of removing the names of each dimension so that the picture appears less cluttered.

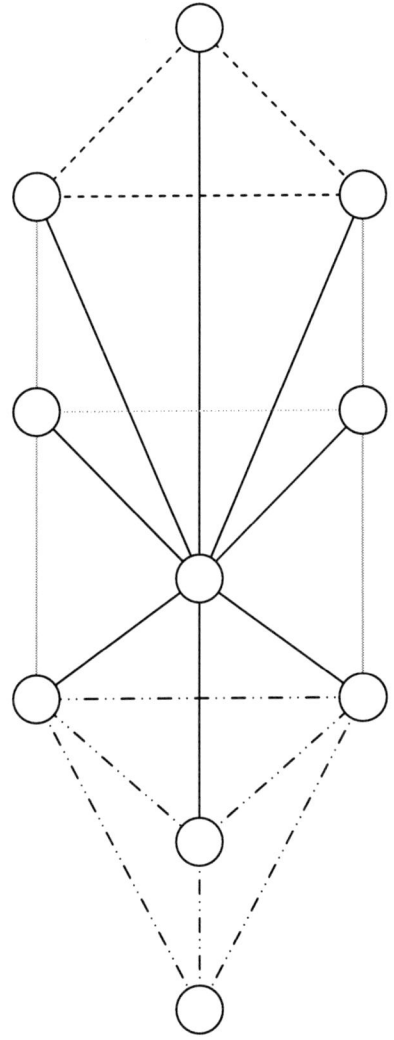

Figure 7: the Complete View of all 10 Dimensions and How They Join

At this point, I strongly suspect many Jewish readers, rabbis and kabbalists (Jewish mystics) in particular, are jumping up and down for joy at seeing a schematic that is line-for-line and ball-for-ball, a perfect representation of the Kabbalistic Tree of Life. Please understand that this was NOT planned when I first put these elements together; rather, it fell naturally out of what was required to get a universe to work and to account for the existence of 'thinking things' in it, as ours does, in much the same way that an omnipresent, omnipotent and omniscient deity fell, naturally, out of my intention to use string theory to solve quantum entanglement; however, the exact likeness to the Tree of Life I find extremely unlikely to be purely coincidental. Moreover, I felt thoroughly vindicated to have re-discovered, using stick-and-ball schematic techniques, one of the most ancient schemas for the universe and to have, in a scientific way, found a reason for each dimension/Sephirah and for each link/path between them.

When I discovered this likeness, I was overjoyed. It was the first meeting of science and religion on exact terms. It took me by surprise for a moment, as I had not expected it in any way but then I realised that this is exactly the sort of thing that should happen when trying to discover the truth about the universe. It wasn't just that religion and science were lining up, they were aligning perfectly and I took this as God's way of demonstrating to me that I was on the right track. I sincerely hope that this likeness can serve as a bond between the Jewish readers of this book and me. Since I began studying the various mysticisms of world religions, the Tree of Life has always stood out in my mind as a perfect way of categorising the various encounters we have both here in space-time and in any afterlife. To have re-discovered it independently in the course of an attempt at describing the universe in a 'theoretical physics' fashion was more than I could have imagined or dreamt of.

But, continuing on, there is much more explaining yet to do. When I viewed this schematic as a whole and realised that it was a perfect duplication of the Kabbalistic Tree of Life, I also noticed another likeness or analogy that I think is just as important to express. The schematic above divides into three components: the Abstract 'Idea' Space, the Consciousness Space and the 4-D Space-Time Continuum. This forms a perfect likeness to the three basic components of a CPU, the Central Processing Unit of computers. The Abstract Idea Space is like the 'Data Space' where certain concepts are pre-defined, the Consciousness Space is like the 'Bus', which is responsible for moving data back-and-forth between the Data Space and the Core Memory, and the 4-D Space-Time Continuum is like the 'Core Memory' area where things are as real as they can be in a computer. The universe, as described above, is a universal CPU and a universal 'thinking thing'. We exist in the space-time continuum - the 'core memory', which is why we experience things as 'real'. It's as real and as tangible as real can be in this universe. Below, in figure 8, I show these divisions of the whole as 'Data', 'Bus' and 'Memory'.

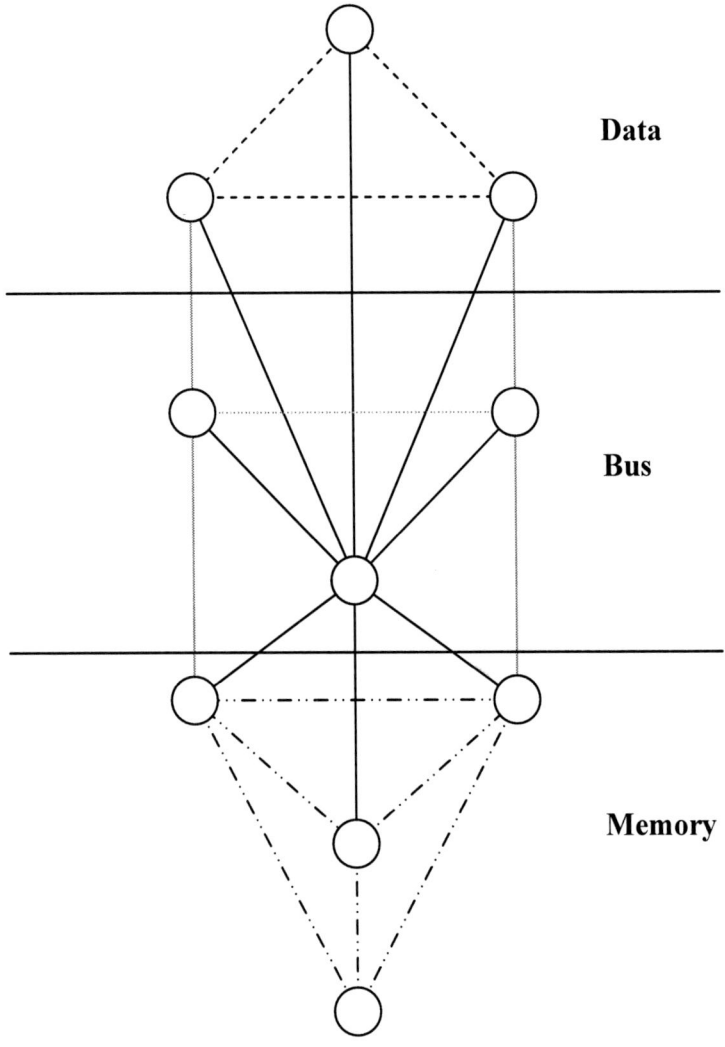

Data

Bus

Memory

Figure 8: The Whole Schematic Showing the Data, Bus and Memory Areas

I think this is a good time to point out that, as the entire universe appears to act like a CPU and that we humans are, in fact, a miniature version of the whole, this demonstrates a couple of known aphorisms and sheds new light on them. Firstly, that we, as microcosms, are created in a "likeness and image" of the whole that is not, in the least, anthropomorphic, but, rather, is a functional likeness. I believe that this is the proper interpretation of "likeness and image" as referred to in Genesis rather than trying to compare man as looking like God in some physical way by having a torso, head, arms and legs. This type of anthropomorphism is sheer folly and misses the point completely. Rather, we humans, as well as any creatures that are 'aware', are constructed in God's likeness and image because we have a physical aspect, a mental aspect and a means via which our mental aspect can access abstract ideas and use those ideas in the physical world. This is what makes us Cartesian 'thinking things' based on the same schema that defines God as THE 'Thinking Thing'; this is the true meaning of 'likeness and image'. Equally, by our human artifice, we have created computers that are 'thinking things', in the very same way, in our likeness and image and in the likeness and image of the universal Creator. Whether or not that was accidental or by a subconscious or latent understanding of what is required to create an 'intelligent' entity, I can't say for sure; although my tendency is towards the latter or even towards a greater understanding of the other pertinent aphorism: "As above, so below".

Perhaps it is instinctive for us as minor creators to develop our intelligent creations after the same intelligent design that stands behind the universe itself. To us, the computer is a microcosm of our macrocosm, yet we are the microcosm to God's macrocosm. This is in perfect keeping with the 'as above, so below' theme and the theme of 'likeness and image', yet nowhere is there any hint of anthropomorphism. It is based solely on the requirements for an intelligent system - a system that is both self-aware and aware of its environment and can interact with its environment and with others of its kind. It also demonstrates - at least, to me - quite clearly that, indeed, Atman is Brahman; but, I'll get back to that later.

Of course, when it comes to God, there are no 'others of its kind'; let me make that perfectly clear. There is no requirement for the true macrocosm to relate to others as it can conform itself into all the various forms of life that we observe in our universe and It is designed, by virtue of the Unifying Dimension existing in 'Consciousness Space', to allow It to perceive in the manner of each of Its creations. The true macrocosm cannot, logically, have others of its kind, as logically there cannot be two entities that are each omnipresent, omnipotent and omniscient. At the highest level, there can only be one; so, God is, most definitely, one of a kind. Anywhere below that level, though, multitudes are allowable.

This schematic gives us the seeds of many concepts that may, at first, not be readily recognised. For example, it clearly has three columns, thus there is a centre, a left and a right. The concepts of left and right can also be viewed as the seed of gender or the origin of any two opposing and/or opposite forces or types and the centre stands as the balance between them. Whilst the whole, obviously, has no gender, it does contain the concept of opposites within its gross structure. And I'm sure that these concepts are defined to the nth level of detail as abstract ideas contained within Its Abstract Idea Space. No doubt, the very concept of gender and of countless opposites would have been discovered by virtue of the whole being self-aware and understanding Its own structure.

Also, there are seven strata or layers in Its structure and now the Hindu readers can jump for joy as the basic schema for the chakras is revealed in the schematic. Also note how the Unifying dimension falls, naturally, at the level of the 'heart chakra', just as one would expect. And, if that isn't enough, the highest Sephirah of the Kabbalistic Tree of Life is called 'Kether', which translates to

'Crown', the name of the highest of the chakras and that which binds us to our highest self - the truth of which is obvious in the schematic. Equally, the base chakra, that which roots us to our place, is time itself. For without time, where else could one root, since time and space are indelibly joined, as we learned from Einstein? The subtleties of the chakras and their linkages are also demonstrated by this schematic, which, in my mind, lends further credence to its representation of truth and the truth of the chakras.

Briefly, above, I mentioned the 'fishing' analogy of how thinking works. If we call this type of thinking, 'Ideation', I feel it better reflects the mechanism, as it is an action that starts from below and reaches up into the area of Abstract Ideas and retrieves them. It is a process that is seemingly driven from our intentions. From the schematic, below, it MAY appear as though Ideation has no access to the 'Form' region of the Idea Space but this is not so; what is true is that our intentions are, in fact, the intentions of the One that are perceived by us to be 'ours'. In every ideation there is an aspect of the Mind of God working, as that, truly, is the only REAL consciousness; our perceptions of our identity must be put into perspective with the veracity of the whole: our individual minds are only 'parallel processors' in the mind of God. Clearly, though, ideation involves all three dimensions of consciousness, as any conscious act must; however, it leans towards the right of the schematic because it depends more on the Individuator than the Integrator, as we are attempting to draw out one thought. After we have 'found' that thought, we THEN incorporate, i.e., integrate, it into our flow of events. In this sense, we use the Individuator as the more active dimension and the Integrator as more passive.

But, there is a different form of consciousness for which we must also account and that is what we term 'Revelation'. Revelation is the placing of a thought within our consciousness from the One. I believe that, in this case, it is more by means of the 'Integrator' dimension that this is achieved. That is, that, in the

71

case of revelation, the Integrator dimension is the more active and the Individuator is the more passive. The passivity of the Individuator is denoted simply by the fact that the revelation occurred to a given individual. In both forms, the Unifying dimension is equally utilised as the whole MUST be conscious of every conscious act. I think it is, perhaps, easier to show this by using the schematic. In the figure below, figure 9, I show the pathways involved in both Ideation, highlighted in grey, and Revelation, highlighted in dots. This shows how the interface between Consciousness Space and the Abstract Idea Space works in both Ideation and Revelation.

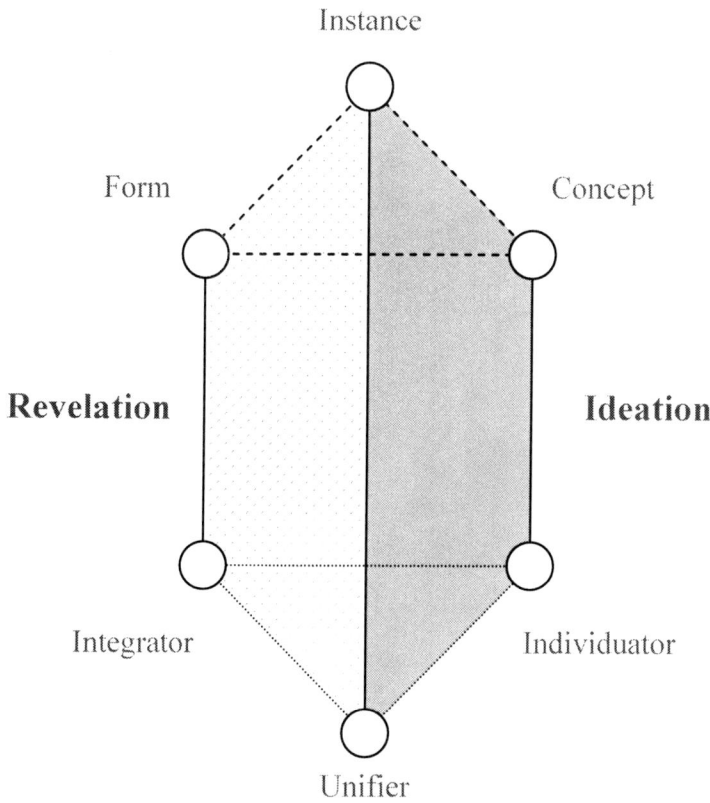

Figure 9: Showing the Paths Involved in both Ideation and Revelation

Now, to be fair to the other interface, the interface between Consciousness Space and the Physical 4-D space-time aspect, I should highlight that, as well. In the figure below, figure 10, the Consciousness/Physical interface and the pathways involved are highlighted in zig-zags.

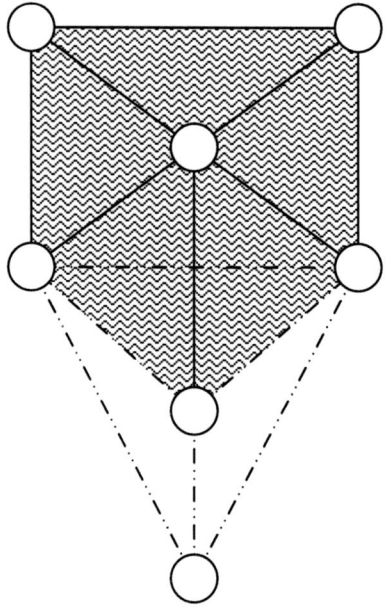

Figure 10: Showing the Paths Involved in the Interface between Consciousness and the Physical

At this point, I think we can, now, show the entire picture, involving all the interfaces and where those interfaces overlap. The overlap, of course, only occurs in Consciousness Space, as it is only within consciousness where awareness of the overlaps is necessary - although, there is, no doubt, an abstract representation of these overlaps within the 'Data' Space. In the figure below, figure 11, I've gone back to showing the CPU-like divisions, as I think that helps clarify why the fields overlap where they do.

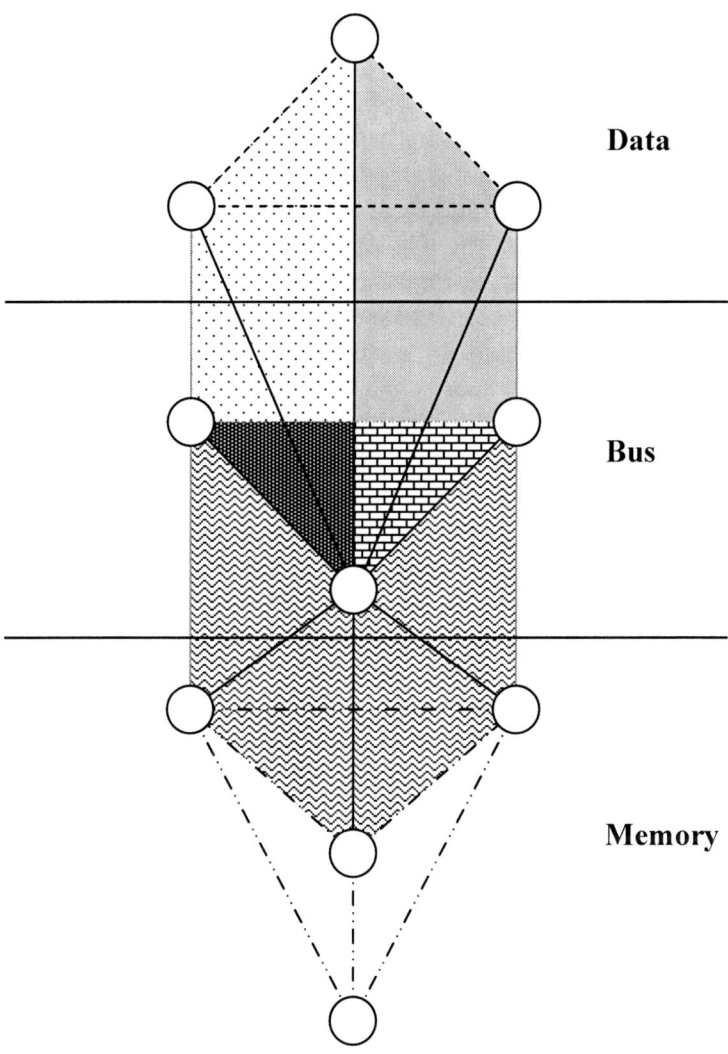

Figure 11: the Complete Schematic Showing All Interfaces

The view in figure 11, above, is the complete view of the whole 'Universal CPU' with all Its interfaces displayed. Remember, of course, that I've given each dimension its own name, although they aren't displayed in this final view in order to not detract from the picture itself. Once the reader has familiarised themselves with this schema, the rest of the book and the various derivations I make will be much easier to relate to. This forms the true crux of this model of physics and it contains many of the metaphysical aspects that support the physics and helps explain the physics behind consciousness, normal thought (Ideation) and revelation as well as where (and a step towards 'how') ideas are stored. It also forms the basis for the macrocosmic view of the universe.

Yet, after all this, I still haven't answered the question posed as the title of this chapter: "What does energy look like"? I've described the 10-dimensional environment in which energy is conserved but have yet to show the implication of that environment on the shape of string-like energy. This massive diversion was required in order to help the reader understand the need for the change in the shape of the underlying strings that act in this universe. Without a fundamental understanding of the three main regions of this universe - the abstract idea space, consciousness space and our space-time continuum - it would be far more difficult to grasp the alteration in topology to the shape of the underlying strings that I put forward. Below, in figure 12, is how I view the shapes of open and closed strings given the environmental constraints of the aforementioned 10-dimensional universe (the 4-dimensional space-time continuum together with the 6-dimensional Calabi-Yau space) and its three-fold division.

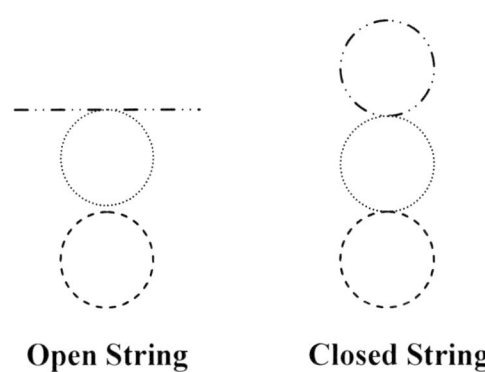

Open String   **Closed String**

Figure 12: New Proposed Shape of Open and Closed Strings

In figure 12 above, I've shown the area pertaining to the 4-D space-time continuum as dashes and dots, the area pertaining to consciousness space in dots and the area pertaining to abstract idea space in dashes. Hopefully, this will help the reader understand and see how the different areas of the same string can be utilised in the various three universal regions. I was tempted to finish that last sentence with 'at the same time' but I thought that might be confusing as time only really affects the part of the string shown in dashes and dots; perhaps 'at once' would be a better closing for the sentence. Nevertheless, the point is that there are three regions to any given string and each 10-D string is able to act in three different ways at any given 'space-time time'.

Whilst it may be true that each aspect of each string is acting in a co-ordinated way - for example, in the case of an open string, the physical aspect is some part of X's brain, the consciousness aspect is a part of X's consciousness and the abstract idea aspect is a part of what X is thinking about - I don't believe this is, by any means, necessary. In fact, for physical objects that do not display any awareness, the consciousness aspect must either lie dormant or, potentially, it could be utilised by some other life-form, which may not even physically exist at the same 'space-time time' as the physical aspect is existing in a non-aware state. Similarly, for a non-aware object, the abstract idea aspect may or may not actually pertain to the concept to which its physical component relates. The ways in which the consciousness and abstract regions of a string are used can only be theorised; they cannot be ascertained for certain due to our inability to see things this small much less be able to see around the dimensional corners into these areas. But I won't let that stop me from theorising.

It may be logical and simpler to keep a certain amount of coherence between these areas and how they are expressed throughout these regions; however, it also may be simpler to utilise the atemporal regions of each string in ways that conserve their energy in the best way possible overall and those particular usages may seem counterintuitive to those of us who are not omniscient, i.e. every human! For example, say that Tom is made up of strings whose 'consciousness aspects' are being used for Susan's consciousness and vice versa. This may cause the universe to put Tom and Susan into a close relationship, as it affords both the physical and the consciousness regions of Tom's component strings to be closer to where (in space-time) they are being used and closer to the physical aspect of Susan's consciousness and vice versa, thus affording, overall, a conservation of energy by saving space-time. Now, this is certainly a radical take on relationships, but it is far from impossible or even implausible. Perhaps it is interwoven relationships like that - at the lowest, quantum level of reality - that cause things like 'love at first sight' or even an uncannily successful business partnership. The universe really is that interconnected at this lowest of levels and this interconnectivity could help explain many otherwise unlikely combinations and/or partnerships.

Now, I believe, I've finally answered the question posed as the title of this chapter and, in the process, given a decent overview of the general structure of the universe - not only at the macroscopic level - but at the microscopic, as well. Plus, I've hinted at how interconnected the universe is and just how twisted and folded and wrapped around itself that it is at the most microscopic of levels. But, from this model, there are still many things yet to be explored in more depth and, in so doing, I hope to impart a greater knowledge of the universe to the reader and I sincerely hope that the reader learns to appreciate and be awed by the sheer majesty of this system in which we are a vital part.

Yes, I said a vital part. This is another philosophical implication of Einstein's Special Relativity. As all spatio-temporal points are already defined in the whole of the space-time continuum, it is fact that everything that exists is, by necessity, vital to the whole. This philosophical implication is further elaborated in the 'Philosophy' section of this book; but, it is worth mentioning it now so as to make the reader understand that everything and everyone around them are equally as vital to the whole as any other. This is a philosophical truth that derives straight from Special Relativity and it is a religious truth to many faiths, as well. Yet again, we have a perfect meeting and marriage between science and religion that can be demonstrated only via String Theory and, truly, only with this particular version of it.

With this chapter, I've detailed what 10 of the dimensions of our String Theory universe do but have completely ignored the other 16 that I stated were, also, a required part of the 26-dimensional String Theory that I said I believed to be the one String Theory that was the most likely. Well, I can no longer afford to ignore those. The next chapter will demonstrate where those dimensions are and what they do; so, dear reader, please continue reading or risk not having an even greater understanding of this incredibly convoluted universe in which we are such a vital part.

What is the Fabric of Space-Time?

Well, it certainly isn't linen. Here, though, we have another opportunity to use a good analogy. It is, with regard to shape and form, a structure not unlike that of the warp of a tapestry. In this case, the 'threads' (actually closed-loop strings, in this case) that form the structural warp for space-time are thinly stretched Higgs bosons, the particles responsible for imparting the gravitational force. Actually, it is only one thinly stretched Higgs boson that is topologically shaped to fill the vast medium through which our space-time expands. The matter of our space-time continuum passes through these warp threads like the weft of a tapestry and the result is our observable universe. It is more accurate to say that the Higgs boson 'warp' extends in all directions, radially, from what was the area of the Big Bang and continues onwards until it reaches the toroidal 'Anti-matter' wall at the end of the dimension through which our space-time is expanding.

The dimension through which our space-time is expanding is, itself, expressed radially from the area where the Big Bang took place and is beyond our ability to view simply because it is, always, around a dimensional corner from our line-of-sight. The reason for this goes back to the chapter where I'd said that the original 26-dimensional string theory model was, in my opinion, the best. One of the reasons I stated was because it predicts the existence of tachyons, which, when they hit that anti-matter wall, condense into a Higgs boson. Tachyons - those faster-than-light particles - begin moving at the speed of light and, as they lose energy, they gain speed; whilst this sounds counterintuitive, it is, nevertheless, true. When they condense, they would leave extremely thin, gossamer-like strands behind them reaching from that anti-matter, outer wall all the way back to their point of origin where Big Bangs occur at the centre of the toroidal medium through which space-time expands and these strands form the long sought after Higgs boson. So, the tachyons moved through the toroidal dimension that extends radially from our 4-D space-time forming the Higgs boson and they would have done this very quickly, as they move faster-than-light speed. This dimension is the 11th dimension of M-Theory. In my case, though, that region contains far more than just one dimension, as it contains all the remaining dimensions, as well.

Now, as I'd said, there are other reasons why I prefer the 26-dimensional string theory and here's another one. Please, don't be put off by the next few statements; they are there mostly for the few theoretical physicists out there who would understand it and the rest of you will be able to grasp the concept by an analogy. In this 26-dimensional model, the 26-dimensional space is governed by Octonion, or 8-dimensional, algebra. One of the advantages to a 26-dimensional model that is governed by 8-dimensional algebra is that the rotational symmetry afforded by the 8-dimensional algebra allows 16 dimensions to be rotated away from the view of the other 10 dimensions. See, I'm talking WAY over most people's heads. Now, you're probably wondering what I mean by rotational symmetry as well as being totally blown away by the thought of 8-dimensional algebra, so, it's time for that analogy. Imagine a perfect cube with one corner pointing towards you. You can easily see that the cube is three dimensional as you are looking at it from a corner. But, if you rotate the cube so that only one face is facing you, all you can see is the face, the height and the width of the cube, the depth of the cube is no longer visible. In three spatial dimensions, a cube allows one dimension of certain 3-dimensional objects to be rotated out of view because of their shape. In EXACTLY the same way, a 26-dimensional object, if shaped correctly and rotated correctly within those 26 dimensions, makes 16 dimensions rotate out of view from the remaining 10. This is a HUGE advantage of the 26-dimensional model, as it means that you can have 16 dimensions that are, from our angle, always invisible.

The dimension that our space-time expands through is one of those 16 dimensions that are invisible to us because it has been rotated out of view. The remaining 15 dimensions are also rotated out of view and I will return to that later. The vast network of the thinly stretched Higgs boson is a structure that exists in this dimension through which our space-time expands. Because it exists in a space that is rotated out of view, our scientists are completely unable to see it, yet gravity operates evenly throughout space-time. This is exactly what you would expect if the Higgs boson, the structure responsible for gravity existed in a dimension that was outside our line of sight and this is exactly our scientific observation. Gravity works perfectly, everywhere, yet there is, seemingly, no structure visible for it to work. Again, string theory accounts for what science observes AND for what science does NOT observe.

Additionally, there are a few other ways of creating tachyons. One process that purportedly creates tachyons is that a ferromagnetic substance, like pure iron, which has been magnetised and raised above the Curie temperature - the temperature above which the magnetic field produced by the magnet can no longer be retained - then cooled again to reduce the temperature below the Curie temperature, can create tachyons. So, this process, as well as other processes I've not discussed, offers us potential opportunities to create tachyons AFTER a Big Bang. I'm sure that, given the age of the universe, there have been times and places where ferromagnetic substances have been subjected to forces that would raise and lower the temperature of that substance above and below its Curie temperature, if so, then, potentially, at least one if not more tachyons would burst forth from that point. I suggest, though, that, when they do burst forth, they do so within the dimension through which our space-time is expanding and, therefore, we do not and cannot detect them. However, these odd tachyon bursts would help form a more grid-like or reticular appearance to the network of Higgs bosonic material that stretches throughout that dimension and the more grid-like that structure is the stronger that structure is. Note that I make the assumption that all tachyons created after the Big Bang would naturally result in additional Higgs bosonic material; this is because the wall of anti-matter firmly exists by that time and, once the tachyons encounter it, they will condense into Higgs bosonic material. These later bursts would appear like nodes on the original Higgs structure and would extend radially in directions away from the node. The only limiting factor to the additional structure is the number of tachyons created during any given tachyon forming process.

But what about these other 15 remaining dimensions, what purpose do they serve? I believe they serve a couple of purposes. They contain some metadata that defines how the various gauge bosons work (this is, perhaps, a point for another lecture or a follow-on book but it is, largely, irrelevant to the general concepts of this theory) and they allow for the expansion of certain other regions of space - in particular, certain regions within 'Consciousness Space', which I will discuss in a later chapter. Exactly how the metadata is defined, I don't know, it's possible that it is managed by a simple twisting and turning of the dimensions (the exact meaning of which would only be comprehended by the whole, which is God) and, as they are rotated out of our view, we would never be able to see them or their structure. Alternatively, these definitions could be implied simply through the spatial relationships these dimensions hold with regard to one another. One of these other 15 dimensions, though, I suspect works together with the dimension through which our space-time expands and allows for the ENTIRE Calabi-Yau space to expand in tandem with our space-time continuum.

That leaves 14 dimensions. This is where I enter the realms of thought that science abhors: religion. In the following chapter where I discuss the shape of Consciousness space, I mention the fact that Islam purports that there are 7 Heavens and 7 Hells and

84

that the Qur'an alludes to the fact that each of them can expand. I believe that the remaining 14 dimensions are the dimensions that allow the individual expansion of each of these 7 Heavens and 7 Hells. By using 14 individual dimensions, this allows for the expansion of each Heaven or Hell without those individual expansions affecting the expansion of any of the other Heavens and Hells. Thus all 16 dimensions that are rotated out of view serve very useful purposes, yet only religion puts forward a potential reason for the existence of most of them.

It's no wonder, then, that theoretical scientists were happy to abandon the first 26-dimensional model of String Theory as it seemed to have more dimensions than was, later, discovered necessary to account for what we observe, plus it purported the existence of tachyons and, as no scientist has ever observed a tachyon, they dismissed the theory in favour of simpler models. Yet the original 26-dimensional model is the only one that not only allows for what we observe in this 4-D space-time, but it also allows for the unseen aspects of the universe that God says It created. So, whilst the 26-dimensional model is, mathematically, the most complicated, it is the simplest model that accounts for both the observable universe and supports, FULLY, what God says God did with respect to Creation! As far as I'm concerned, it handles these subtleties in the most elegant manner possible, which is, again, something one would expect from an omniscient, omniprehensive and omnisentient deity.

As far as where the other 15 dimensions that are rotated out of our view exist, it's possible that they are entwined in a line-like fashion around the strands of the Higgs boson making the whole 'warp' of the fabric of space-time look a bit like a cable with the Higgs boson in the centre and the other 15 dimensions heliacally wrapped around it in a similar way to how DNA is wrapped heliacally around histone molecules in the chromosomes of each of our cells. Perhaps the metadata that those 15 dimensions hold is analogous to the histone code of epigenetics, which is responsible for the varying ways that genes are expressed. A similar correlation could explain why some spatio-temporal events are non-repeatable, simply because the bosonic metadata doesn't allow for it.

By the way, another classic reason that scientists will never see a tachyon is that, if they increase their velocity at the same rate that space expands or greater than that rate, they would have all passed us by now. In fact, although I won't bore you with the mathematical details, the universe would have only needed to have produced tachyons for the first 100 million years - roughly 7.5 percent of the age of the known universe - and, even then, they would have all passed us by now, discounting the fact that tachyons, when produced, are rotated OUT of our view, which also prevents them from being seen yet perfectly allows them to serve their purpose. In fact, if that network of 'warp' were visible from our vantage point, it would cloud up the universe so much that it would be almost impossible to see anything else; so, naturally, a God that is all-wise would keep them out of view so that Its creations that are conscious and have a visual sensing ability, wouldn't be blinded by the very structure that holds up the universe that we CAN see. Indeed, The Qur'an alludes to this when it mentions that Allah created the world with unseen pillars, which is about the best analogy you could give to 7th Century humankind!

One more thing that the 'fabric of the universe' touches upon is the fact that Edwin Hubble discovered in 1929, i.e., the fact that space itself is expanding. This fact is a huge revelation, as it has a direct impact on Einstein's Special Relativity. Special Relativity states that both space and time are relative to the constant speed of light. However, if the speed of light appears to be constant and space is expanding through a medium, then the photons MUST be accelerating in order to, relativistically, appear to be moving at a constant speed!

If the distance between A and B is expanding due to the expansion of space through the dimension through which it is expanding, then photons must accelerate through all 5 of those dimensions in order to appear to be moving at a constant rate in our 4-D universe of Special Relativity. In fact, it would have to accelerate at the same rate that space-time is expanding, in other words, the inverse of the Hubble Constant where the first two terms are reversed in order to show a temporal contraction, thus: $(1/H°)$ per second per kilometre per Megaparsec. Which leaves the question, "What is governing the rate of expansion of space?" I believe this, very well, may have been the 'cosmological constant' that Einstein first envisaged in his first draft of General Relativity. He was on the right track but, the concept OF Special Relativity soon made him realise that, relativistically, this cosmological constant plays no role with respect to those of us inside the 4-d space-time continuum. Rather, it only becomes obvious when one tries to 'view' the space-time continuum from outside - from that 5th spatial dimension (the 11th of M-Theory) through which our space-time is expanding.

Plus, we have noted that clocks that are moving relative to clocks that are not moving experience time 'dilation'. Time dilation stands opposed to what I would have expected to find for time given the spatial expansion Hubble discovered. From Hubble's discovery of expanding space, I would have thought that the passage of time would decrease or slow (a temporal contraction!) at a rate inverse to the spatial expansion and that THAT would account for C appearing as a constant. However, the time dilation aspect of Special Relativity stands as evidence against such an inverse temporal contraction leaving me with no choice but to accept that the photons are accelerating, thus moving faster than light speed at a rate that is equal to the inverse of the Hubble Constant times C/v where v is the velocity of the moving object, i.e., "$((1/H°)C/v)$ seconds per kilometre per Megaparsec". Accepting this, it forces Science to acknowledge that C is not a normal 2-vector, that is, an element that indicates speed in a particular direction, but that C is, in reality, a 3-vector that indicates speed in a particular direction accelerating at a particular rate. It also forces us to acknowledge that the system in which we are living is, with respect to C, not a 4-D Minkowski space but a 5-D space where four dimensions are spatial and one temporal and, thus, a triumph for M-Theory; yet, here, it is a triumph for M-Theory expressed across 26 dimensions rather than only 11.

I suspect, although, at this time I haven't 'done the work' (my reason for that is below!) is that, with respect to one another, our 4-D space-time of matter and the 'wall of anti-matter' at the boundary edge of the dimension through which we are expanding - the 5th dimension alluded to above - are magnetic monopoles with respect to one another, as the matter has, overall, a positive charge and the anti-matter, overall, has a negative charge (or vice versa, the main point is that they are opposite magnetic monopoles with respect to one another). As the anti-matter 'wall' is static but the 4-D space-time is moving (via its spatial expansion), the opposite charges would attract one another and THAT is the driving force behind the expansion of space, contraction of time and the required acceleration of photons making the speed of light to appear, in 4 dimensions, to be a constant. I believe that the field force of this magnetic attraction IS the cosmological constant to which all things are relative and I believe that, once that value has been calculated, not only will we know when the matter of this 4-D space-time will encounter the anti-matter wall, but the theory that I've just described, regarding the FACT that C is NOT a constant - a theory we could call "Extra-Special Relativity" - will be complete. With that, we will finally have a consistent theory that explains where the Higgs boson is (rotated out of our view), what caused it (the presence of tachyons and their condensation into the Higgs boson), what allows for the expansion of 7 Heavens and 7 Hells and what causes the expansion of space itself. The down-side to this is that we will also be able to calculate, with reasonable precision, how much longer space-time has before it smashes against the wall and total annihilation results and the entire lot of matter/anti-matter energy wraps back around to start the whole process over again with another Big Bang-to-Annihilation sequence.

I strongly suspect that many of us may not want to know when the end is coming, although I admit that several WOULD want to know. I strongly charge the person who wants to calculate that time to keep it to themselves in order to not cause a world-wide panic. That is the main reason why I haven't done the work to calculate that missing 'cosmological constant', which would be the attracting force between the total mass of our (spherical) 4-D space-time acting as a magnetic monopole being drawn to a toroidal wall of anti-matter of the same mass acting as the opposing magnetic monopole. Additionally, I believe that this is the main reason why scientists haven't observed a magnetic monopole within our 4-D space-time. It's because there are only two monopoles: one is our expanding 4-D space-time and the other is the anti-matter wall - which is, in truth, a component of space-time, albeit invisible from our perspective - neither of which can be seen in total. We can't see the entirety of the expanding, matter-containing aspect of the space-time continuum because we can't get outside of it nor do we have the technology, yet, to even see all the way back to half its diameter (the point of the Big Bang); we can't see the wall of anti-matter because it lies at the very end of our future and we can't even see one nanosecond into the future by scientific means.

Nevertheless, if someone does perform the integral calculus required to discover this magnetic 'cosmological constant', we will be able to calculate from that the likely end-time of our expanding 4-D space-time. Again, I stress the importance of not causing a panic. The time could be within 50 years or it could be 50,000 years or more. To divulge this information at any time could result in a panic right before the end and this, I believe, would be a tragic thing. To cause the kind of panic that would, in all likelihood, cause people to act in ways they otherwise might not

- because they know 'the end is nigh' - and, in so doing, potentially jeopardise how they spend the remainder of their eternal existence, is not something of which I want any part. Although by revealing what I have already revealed, I'm reasonably sure that some physicist will want to perform the calculation and all that panic will be on that person's head; but, it won't be on mine.

What Does Consciousness Space Look Like?

Here is a simple question that will lead us to the answers to many other questions that are both physical, as in pertaining to the physics of consciousness, and metaphysical because consciousness is the area in which we encounter the most mystical and magical as well as the most mundane of all of our experiences. In fact, it is the only area in which we encounter and recognise any experience whatsoever, which is why it is so vital to understand it. Now, so that detractors won't call me blatantly foolish, I will state, plainly, that what follows is both speculative and, as yet, unproven; however, that does not mean that I'm incorrect. All science begins with hypotheses and then sets out to prove or disprove them. There may be, in the future, some means of testing some aspects of my proposal but, at the present, I use inference based on experience, which, when it comes to consciousness, is the only real way forward at the moment. Of course, I also accept my own model as the basis for my speculations and I can think of no other individual - scientist or philosopher - who wouldn't do and hasn't done the same. There, consider the necessary, technical apology acceded... now, on to the business at hand.

In a nutshell, the answer is "very much like a fine-toothed comb", but that is only the grossest of analogies. The 'teeth' of the comb are more like very thin sheets of paper, thus the width of the comb is defined by the total amount of individual 2-dimensional aspects of consciousness and the teeth of the comb are our individual 2-dimensional consciousnesses - this would, naturally, include the 2-dimensional teeth that represent ANY form of awareness. Earlier, I've contended that we each have a 2-

dimensional 'slice' of what is, overall, a 3-dimensional consciousness space. I've also described why we need two dimensions to our consciousness (one to provide us with the ability to apprehend a moment and the other to integrate those apprehended moments into a flowing whole) and I believe this is correct. Additionally, I've purported that these dimensions must be beyond the 4-dimensional space-time as they are, in a sense, a different type of dimension than either true spatial or true temporal dimensions. The third dimension of consciousness space is used by the whole (the one, true, single entity that exists - all it God, Allah, YHVH, Brahman or any number of other names) in order to integrate all of those individual slices into one contiguous 'loaf' of consciousness. It is by maintaining and retaining that 3-dimensional loaf of consciousness in atemporal space that grants the universal entity the attribute of omniscience. By the way, I use the 'slice-and-loaf' analogy because I believe it helps the reader grasp more easily the relationship between the 2-D and 3-D aspects of consciousness and not because I believe that consciousness really looks like a loaf of bread.

Above, I stated that the best overall analogy is that of a fine-toothed comb. In fact, I repeat it may best be said that the teeth of this comb are as thin as the thin can be; that is, the slices are as thin as they need to be in order to have as many as is required by the whole, although, the thickness of each slice/sheet of paper/tooth may vary and this could account for a variance in the level of consciousness expressed. The difference between the fine-toothed comb and the loaf analogy is that, in the comb, there is an area where there are no teeth, i.e., no slices - an area that is completely contiguous and continuous with itself. There is a perfectly legitimate reason for requiring such an area and the main reason is simply because it's possible. God, in my view, is - and must be - a glutton for experience and, because It is the only thing that experiences everything, It gains all those experiences by making them occur.

The non-toothed area serves as a place for consciousness-scapes, like our dreamscapes, where we play out our dream experiences, and the potential for an actual Heaven and Hell where we exist in a region of reward or punishment. Thus, within the non-toothed area, there may be - and probably are, because it's possible - several divisions so that we can enter and/or leave these areas and encounter the things that can only be encountered in such an area.

This topology for consciousness allows for and gives firm physical reasons for dream experiences and the potential for eternal rewards and/or punishments. Note that any time (that is, time as passed in space-time) spent in atemporal space would seem eternal for the individual experiencing in atemporal space. Perhaps it is the very fact that we maintain a physical interface to space-time when we dream that we experience time in dreams at all, although any dreamer can verify that time spent dreaming does NOT pass at the same rate as waking time and that it can vary from dream to dream.

Also, let us not forget that humans aren't the only dreamers in our universe. It is well known that dogs enter REM sleep and countless dog owners have noticed when their dogs were dreaming. Unfortunately, we humans aren't bright enough, yet, to converse with dogs to understand the complexity of their dreams; it's safe to say they do, at least, dream. Nor have dogs invented a means of written communication, but I think it's safe to say that they can communicate with others of their kind and, probably, qualitatively better than, say, slugs do but, perhaps, less than we humans do.

I believe that, in animals, it is the structure of their nervous systems that allows the level of consciousness that the animal can display. In the Penrose-Hameroff Quantum Consciousness theory, there are structures of tubulin in the nervous system that allow a certain quantum flux that they - Penrose and Hameroff - believe could account for consciousness. I believe that the Penrose-

Hameroff flux is the physical interface TO consciousness space rather than consciousness itself. The tubulin structures, as well as forming the natural, skeletal framework for the neuron (a neuron is a 'nerve cell'), are close enough together to allow for information to flow between the Calabi-Yau space and our 4-D space-time and it is that flow - that quantum flux - that forms the interface between our nervous system in space-time and our actual consciousness in the Calabi-Yau space.

I realise this is highly technical and requires a certain understanding of cellular microbiology; but, if I don't clarify myself here to some extent, scientists - including Profs. Penrose and Hameroff - will have good reason, metaphorically, to bite my head off. With that in mind, please note that when I say 'tubulin structures' I do not intend that to be interpreted as 'tubulin strands' but the combination of α-tubulin and β-tubulin that form the tubulin dimers that, when strung together, form the long polymer strands of tubulin. These dimers have magnetic polarity; and, whilst we know that the polarity of these dimers controls the overall polarity of the strand, it could well be that this polarity could also act as an informational 'bit'. That is, if the quantum flux that goes into the Calabi-Yau space comes from the negative side, the result is 'False' and if the quantum flux that goes into the Calabi-Yau space comes from the positive side, the result is 'True'. In this way, information can be passed to and from the Consciousness Space through these structures. We know that neural axons of white matter grow towards the dendrites of grey matter neurons BECAUSE of the growth and polarity of tubulin. This is highly suggestive that the neural pathways between white and grey matter represent the interface to conscious thoughts or memories and that they are, in fact, founded in the tubulin arrangement that must come first. So, as I'd said, I believe what we see in a Functional MRI is not consciousness but only the electromagnetic aspect of the interface to consciousness; however, if I delve any deeper into the actual microbiological

94

aspects of this, it would be way beyond the comprehension of most readers and I do not want to hinder most readers. Perhaps a follow-up lecture or paper discussing this further would be more appropriate.

If my thinking is correct in this regard, then it is the relative complexity of the nervous system in animals that helps define the quality of that animal's consciousness. In plants, their awareness is purely chemical, yet there is, most definitely a level of awareness in them. And let us not discount this, as the various hormones that we have are merely an animal example of this type of chemical awareness. But, while I can imagine that a dog may have the leisure time and the raw ability to muse about its own existence, it doesn't have the ability to prove that to our satisfaction and I seriously doubt that the yew tree - no matter how long-lived it may be - has the capability, through chemical intelligence alone, to muse about its existence. There are levels of awareness and the complexity of the life-form determines qualitatively and quantitatively, the level of awareness/consciousness that the life-form is able to perceive and exhibit.

I propose that our consciousness is, in essence, a coherent, 2-dimensional field of conscious energy that is comprised of several of the Consciousness-Space aspects of the underlying strings of energy. And, as our individual field of consciousness actually exists in atemporal space, it is, therefore, eternal. This is the 'reason' for life after death and the reason for any potential life before birth. This theory does not discount the possibility for re-incarnation; rather, its physics supports the possibility. Because of that, the logic of omnipotence would demand that there are a few different types of incarnations:

1) a single incarnation that was agreed upon between God and the individual involved: a willing incarnation
2) a single incarnation that was forced on the individual by God: an unwilling incarnation

3) a re-incarnating individual where the aspect of re-incarnating was agreed upon between God and the individual involved: a willing re-incarnation
4) a re-incarnating individual that was forced to re-incarnate by God: an unwilling re-incarnation

Each of these different styles of incarnation, as they are possible, must be explored by an omnipotent entity and this is completely logical. Which of these methodologies is the most prevalent is unknown and, I believe, unknowable without a direct revelation stating the answer. At least, I can think of no reason as to why one or more of these methods would be preferred by the Divine Entity. I can see reasons of 'conservation of energy' behind re-incarnation but, equally, I can see 'conservation of coherent experience' and an emphasis on 'personal responsibility whilst incarnate' in single incarnations; so there are arguments for either side. With respect to force, again, there is little to direct us towards what methodology is most prevalent; although I can imagine a scenario where, for example, God requires an 'Adolf Hitler' or 'Charles Manson', gathers together a group of 'potential individuals' and asks, "Any volunteers?" The likelihood of anyone volunteering for such 'apparent' evil, if given the choice and the knowledge of what punishment might befall them, would preclude most individuals from volunteering, thus necessitating force. I personally believe that part of Hitler's own personal Hell is the knowledge that he didn't kill 6 million Jews; rather, he saved their souls and each of them was welcomed into Paradise. Because his intention was to kill, the equal and opposite reaction resulted: he saved them. Perhaps, also, he must experience how each of them died; I don't know. I do know, though, that justice will be done and mercy will prevail.

But, this scenario also leads us to the question of 'necessary evil'. Someone MUST be Satan or Hitler or Stalin or any number of other evil entities and, I believe, it is most likely that these entities are, more than likely, forced into these roles, as no well-intentioned 'potential individual' would choose such a role if offered the choice. Then again, perhaps this very reason lends the omnipotent Creator the rationale to create 'potential individuals' that are evil-intentioned. What it takes to BE such an entity is either a sincere evil power streak manifested in the individual BY God for that very reason or an incredibly loyal individual that understands that God requires what might appear to be 'evil' to millions of humans but, inevitably leads to some beneficial end, as that is what God must do if possible - turn an evil into a good (I'll explain the reason for that later!). As a classic example of what I mean by that level of loyalty I'll use a story from the Qur'an.

The Qur'an alludes to the fact that it was out of sheer loyalty to God that Iblis (the entity that is the most evil according to Islam), the leader of the jinn (note that Iblis is a jinn and not a 'fallen angel'), would not bow down to Adam as God commanded all angels and jinn to do. Iblis was SO loyal to God that he - even when commanded to do so by the Almighty and knowing full well the punishment was to be eventually damned - could not bring himself to bow down to Adam or anything less than God. This 'trick' of asking the angels and jinn to bow down to Adam uncovered Iblis' loyalty and, henceforth, it was bestowed upon Iblis the duty to perform the tasks deemed 'most evil'. It was not necessarily his choice, but Iblis took up the task and only asked to be allowed the respite of being allowed to exist without punishment and not to be punished until the 'Last Day'. This, of course, God wisely and graciously granted, purely because it allowed God to use Iblis in the role of Supreme Evil up to the very last moment.

The topology for Consciousness Space that I propose offers an area where such a story could take place, an area that is common, usable and accessible to conscious entities like jinn, angels and the presence of the Almighty Itself; this would be an area in the non-toothed region of the comb. As my model of physics has to account for both physical reality as we observe it in our 4-D space-time (and this would include dreams) as well as religious concepts, there must be common areas where the little slices or sheets of our 2-D consciousnesses can be brought together to have common experiences as in the example above. There are several of these common areas put forward by different faiths.

Christianity purports a Heaven and a Hell. Heaven is a place where the souls (which are viewed as coherent fields of consciousness in my theory) enjoy an eternal reward for their beneficent acts whilst bound to a body in this 4-D space-time continuum. Hell is a place where souls are punished for varying lengths of 'time' up to and including eternal punishment for their evil acts whilst bound to a body in this 4-D space-time continuum. Judaism, in its more mystical texts, alludes to the existence of varying layers for both Heaven and Hell and that the rewards or punishments are dependant upon the level within either Heaven or Hell in which the soul exists. Islam states firmly that there are 7 Heavens and 7 Hells and the rewards and punishments vary according to the level of Heaven or Hell in which the soul is cast.

None of these different views conflict with one another. In fact, they differ only in the degree of detail they put forward. The most detailed account is from Islam and it matches very closely with earlier Jewish mystical texts' views. This is hardly surprising as all three of these faiths purport to worship the same deity and to have their revelation from that same deity, albeit in different ways. The Qur'an, though, has another statement in it that is both metaphysical and truly physical with respect to the Consciousness Space. There is a statement (In Qaf, Surah 50:30) in the Qur'an that states, "One day We will ask Hell, 'Are you filled to the full?' it will say, 'Are there any more?'" The physical implication of this passage is that Hell can expand to allow for more souls. Although it is not specifically stated that Heaven also has this ability, I believe that it is reasonable to assume it would have to in order to be consistent. But what mechanism allows for this expansion? This mechanism is, of course, the existence of 14 of the 16 dimensions that have been rotated out of our view due to the rotational symmetries afforded by 8-dimensional algebra in a 26-dimensional space as I've outlined, above, in the chapter, "What is the Fabric of Space-Time?"

I believe there is also an area of 'Consciousness Space' that is our dreamscape, the place where we play out our dreams. This area needn't be expandable; it just needs to be large enough to handle the space required for our dreams to take place - of course, that includes the dreams of ALL creatures that have the ability to dream. I don't, of necessity, limit that to humans alone. I've mentioned, above, that dogs dream and, I suspect, several other animals retain this ability; neither will I nor can I logically rule out extra-terrestrial entities from having this ability. Nevertheless, I still believe that the dreamscape itself does not require the ability to expand; rather, it just needs to be vast enough to process all the dreams that dreamers have. But there is a question I have about the dreamscape. Is there just one that we share or do we each have our own section of the dreamscape?

One way to discover the answer to the question of whether or not a shared dreamscape exists would be to take a group of lucid dreamers and have them perform this test. Each dreamer must be able to lucidly dream about the same place, for example, Central Park in New York City. All the dreamers must be familiar enough with the place that they dream about such that they can go to a particular place within it. The test involves having one dreamer go to an agreed upon place and write a note - the text of which is known only to the original dreamer - and leave it there for the next dreamer to find. If the next dreamer finds the note and is able to read it and read it verbatim, then we have proven that the dreamscape is, indeed, a shared space. If that could be proven, the dreamscape could be used to pass covert messages from person to person. Searching reality and searching cyberspace are tricky enough but also having to search the entire dreamscape for a message, which could be anywhere, would be virtually impossible. The only way, though, to find out if we do share a dreamscape is to have lucid dreamers perform this kind of test.

Even if our individual dreamscape is a section or partition of an overall larger dreamscape, we still may be able to grant authority to another to enter our area. In fact, when we dream about others, they are either our own inventions, based on our memories of those people, or they are, in some respect, a real aspect of those other people. Personally, I have had dreams about people I never met and it was only later in life that I discovered who the individual was who spoke to me in the dream. One such individual was Helena Petrovna Blavatsky, the founder of The Theosophical Society. She died long before I was born and I dreamt of her before I had heard of her. In fact, it wasn't until I saw a picture of her later that I realised that it was her in my dream and I saw the picture about 2.5 years after I had the dream. The woman in my dream spoke with what I felt was a slightly Slavic accent yet speaking English well and it turned out that Helena was from Moldova and she was fluent in English. I could not have a memory of her since we were not contemporary nor did I know anything about her when I had the dream, so how could I have known to give her an accent that was appropriate? The answer could be that it really was some aspect of HPB that I encountered in the dreamscape.

I believe almost all of us have, at some time, dreamt of someone who has died, although it is rare that we dream of someone we have never met, unless that individual is a close relative, like a grandparent, that we wished we had met. As those "'fields of consciousness' that we truly are" are freer than those of us who are still bound to a certain material aspect, it's perfectly reasonable to believe that they are free enough to enter either a shared dreamscape or enter any dreamscape that they have been given permission to enter. I believe that permission can be granted by either us or by God. God may well have certain reasons for us to have certain dreams and in certain dreams God may well have certain reasons for us to encounter particular individuals in our dreams. If this is so, then we will only know the truth of it once we meet our Creator. With respect to my encounter with Helena Blavatsky, I will not go into detail in this book, as it was a private matter that involved the life and death of a friend. I will say nothing further about it other than to say that what she predicted was exactly correct and was predicted about 10 years prior to the event. Helena, as I later discovered, was very much a believer in 'the spirit world'. I strongly suspect that she is continuing her research in the next life as best she can.

I believe that my dream encounter with Helena Blavatsky was one of those scenarios where, if we have our own personal dreamscape, then God gave her permission to enter mine; if there exists a shared dreamscape, then she was free to do so anyway. Yet how do we, if our dreamscape is personal, grant permission for others to enter it? I believe this can be done consciously by making an agreement with God whilst conscious; this would certainly be the way a lucid dreamer would do it. But, of course, you also need the cooperation of the other individual if there is to be a meeting in the dreamscape. This could only be done by God, which is why we have to make our agreement with God. God, then, takes up this request with the so-named individual and an agreement is either made or not made. I would say that based on my own experience - as a lucid dreamer - that it is far easier to get the cooperation of those who have passed on than to try to dream of a living individual.

Living individuals, i.e., those still bound to the material, at some point must make an agreement with God to confront you in the dreamscape. This kind of confrontation with God could (and most likely would) come as a great surprise and the request could tend to shock a living individual who is not spiritually aware enough to allow for such encounters to be considered normal. For example, if I wanted to dream about meeting Emma Watson (Forgive me, Emma, for choosing you relatively randomly! Nor do I wish to imply that you are less than spiritually aware!!), the actress who played Hermione Granger in the 'Harry Potter' films, she might find it stalker-like that someone she has never met would want to dream about her. In fact, she might think it very creepy even if my interest was just to ask her about what she planned to do now that she's freed from her Harry Potter contract or what she would like to do in the future with regard to films or anything. Plus, I happen to think that Emma would be perfect to play the main character in a fiction book I'll be writing after this and I would want to know if she would be interested (THIS, Emma, is why I said 'relatively randomly' above) in playing that lead role if the book ever made it to film because I know she would be perfect for it. However, if I wished to dream about Marilyn Monroe, I believe the task would be far easier. Those who have passed on know far more about reality as a whole than those of us still bound to the material world and, so, would be more likely to agree to meet anyone, even a fan from their relative future.

Another aspect of 'Consciousness Space' that we haven't discussed is the difference between mind, spirit and soul. In my way of thinking, the soul is your total consciousness - the complete totality of the part of you which retains your experiences including all your dreams and any time spent in potential previous (or relative future) lives and any part where you have spent time in Heaven or Hell. The mind is a functional mechanism that the soul uses in order to process thoughts; it is, if you will, an organ of the soul. The mind fetches into the Abstract Space where ideas are retained and retrieves concepts and returns them to the soul. In a living individual bound to the 4-D space-time, the concepts returned are returned to a soul that has an interface to the physical world and can act in the physical world based on those thoughts. The unbound soul is, essentially, free to act within whatever part of

Consciousness Space in which they happen to exist. Spirit is the type of energy that comprises both the soul and the mind.

In Figure 12, above, the 'spiritual energy' is that part of the string shown in single dots and single dashes; in other words, it is everything that is not strictly physical. The soul is mostly comprised of the single dotted area of the string, that is, the middle portion but the mind is a combination of both the single dotted and single dashed, for it must have access to the Abstract Space. Therefore, roughly two-thirds of the energy that comprises us is spiritual. Now this does NOT account for dark energy. Dark energy is purely a physical thing that I will explain in a later chapter. The spiritual energy is the energy that is contained within the Calabi-Yau space. Therefore, our souls are, in effect, a 5-dimensional object, although they are mostly a 2-dimensional object of pure consciousness energy, there is the mind that is 5-dimensional, as it must have access to our 2-D consciousness as well as the entire area of Abstract Space in order to extract and/or process information. When our souls are bound to the physical world, the entire entity is a 9-dimensional object, as it is comprised of a 5-dimensional mind/soul combination made of spiritual energy joined to a 4-dimensional spatio-temporal aspect made of physical energy.

What Does the Abstract Idea Space Look Like?

Although I've touched on the general concept of the Abstract Idea Space and how it contains the abstract ideas to which we all have access when we think, I have not deeply discussed the methodology that is used to store those abstract ideas. And we know enough about encryption to know that data can be complex and while, you may know a method for how to discover a password, for example, it would take longer than you live to actually determine it. Once again, I shall cross that terrible boundary called speculation. This has never prevented me before from doing just that; and, for the same reason, I shall attempt to give the reader an idea of how the concepts MIGHT be stored. If we do not first speculate, we can never know about anything. So, I speculate.

Firstly, remember that the space itself is a 3-dimensional space that is, topologically, pointless. In other words, its area is perfectly smooth affording it the ability to store an infinite amount of data without ever running out of space. The topology is 'without points' and, as any 'point' in space could be no smaller than the Planck-length, it must be an area that is 'without points', as that is the only topology that could afford it enough space to contain what it must contain. It must be, somehow, smooth across all three dimensions. I named those dimensions as: Concept, Form and Instance. I did that in order to separate the different aspects of an abstract concept. These three simple classes form a way to categorise abstract ideas and categorisation is what lies as a substrata - a layer that lies underneath and behind the actual data -

for any formal logic or metadata. In order to have logic, you must have certain categories understood first.

The dimension called 'Concept' is, by far, the most general. It contains the barest layer of information. Such things as 'individual', 'container', 'system', 'operator', 'function' and 'process' are the types of concepts that are encoded in this dimension. These are, for the most part, the various categories that specific forms are dependent upon and that the various forms of which are specific types. The methodology that is used to encode these concepts is, I'm sorry to say, beyond my capacity to state with any confidence; however, I suspect that it is likely to be in the form of slight topological variances and that these are decoded by the mind of God, which is a 6-dimensional entity that comprises all other minds as well as its own unique area. This, in turn, requires that our minds contain a decoding element to them that allow us to make sense of the way in which this data is stored. This would be true for the encoding and decoding of each dimension's coding structure. Again, how the mind does this is beyond my comprehension but that's not to say that it couldn't be worked out in the future. The 300,000 various functions that our livers perform were, at one time, completely beyond our comprehension but are, now, for the most part completely understood; but, it takes time and more minds and hard work to discover and/or uncover such things. For the answers to some of these questions it will take our best logicians and mathematicians. Omniscience would be, naturally, difficult to comprehend, wouldn't it?

The dimension called 'Form' is the most elaborate of all the dimensions in the Abstract Idea Space. For it is here that everything is defined to its fullest. It must inherit attributes from the 'Concept' dimension but also adds its own details that are further refinements of the underlying concepts. Plus, its definitions must also have, within them, inherent relationships to other forms and, perhaps, other concepts in order to fully define

106

them. When I say 'fully define', I mean all the way down to the quantum level. This level of definition is not, necessarily, used by us, but it is absolutely vital in order for God to maintain omniscience and, to a lesser extent, omnipresence. Without this level of definition, it would be impossible to have perfect knowledge and perfect understanding of that knowledge; and, with regard to omnipresence, it must take an enormous amount of capacity for data to know when and where all things exist down to the quantum level. For God, though, this is easy, as it is as simple as folding hands.

Our minds only need to grasp the grosser definitions in order to work for our level of intelligence and understanding. For example, we don't need to know the exact quantum state of the chair in front of us in order for us to understand enough about it to know that the chair is there and is either already being used or can be used by us. In fact, by the time we were to understand the true quantum-level state of the chair, it would have changed if for no other reason than that time has moved on and the chair is no longer in the exact same place it was when we began our investigation. If you doubt this, remember that, if the chair is in your house, your house is on Earth and Earth is rotating on its axis and revolving around the Sun whilst the entire solar system is revolving around the galaxy and the galaxy is moving through space and space itself is expanding through another dimension. All of these movements change the spatio-temporal coordinates of the chair and each quantum particle of the chair and those quantum particles must always be completely defined in the mind of God and that includes, by necessity, their placement within space-time.

The dimension called 'Instance' defines how the various forms exist. I could have named this dimension 'Existence'; but, I chose 'Instance' because it is closer to the computer analogy for the entire 10-dimensional object.   In computer terminology, 'instantiation' is the creation of an object that has already been defined by code. It is a very important aspect of Object-Oriented Programming (OOP). First, the programmer must define an 'object'; only then can he/she create it via the process of instantiation. This is how a 'window' pops up in Microsoft Windows, even if the 'window' is only an informational or error message. The size and shape of the window must be defined and where the window will appear on your screen and any text that appears on the window as well as any clickable buttons. In this example, the 'Concept' is 'Window' and the 'Form' is the entire definition of size, displacement relative to the screen, the message to be displayed in the window and the existence and functions of any buttons that may appear on it. When a computer runs across a scenario that demands that this particular window is created, it instantiates this window: The Microsoft Windows Operating System creates the window, as defined, and displays it to the user and the user responds to it, hopefully, in an intelligent manner. But, sometimes, the computer does say "no". And God, too, is allowed to say "no". God says, "No, you can't have triangles with 4 equal sides today... or, any day!"

So this 'Instance' dimension is concerned with whether or not the 'Form' exists in the physical 4-D world or whether or not it exists in one or more dreamscapes, Heaven or Hell and, for ideas that - seem to us - to come into being, it defines when these forms come into being and to whom or what (some non-humans do make use of tools and pass that knowledge down as culture). It is in this dimension where growth of consciousness is recorded. When an idea that has already been defined is 'thought of' by a human (or any other 'thinking thing'), it enters a new level of existence; it moves from the mind of God alone, which contains all abstract ideas, and shares itself with the mind of the individual that had the idea. That individual may, depending upon that idea, decide to create an instance of that idea. If this is done, then the idea enters a new level of existence where it is fully physical.

The creation of the wheel is a perfect example of this. It may come as no surprise to discover that God knew about the wheel long before it entered the mind of a human. But, probably somewhere back in ancient Iraq, in the Fertile Crescent some 4,000 years ago, some human had the idea of a wheel come into his/her mind. I suspect that it may have been a potter, as it seems that pottery and potter's wheels were in use before it was discovered that, if you turn the wheel 90 degrees, you can use it for rolling things. Once the idea was had, there have been several different types of wheel created since its inception: wheels for carts, wheels with spokes, wheels with cogs (thus the instantiation of a 'gear'), and wheels with tyres around them. All these varying types of wheel had always been in the mind of God but, slowly but surely, they moved out of strictly God's mind and into the minds of humans. This process, although it only changes the definition of a 'Form' in one dimension, the 'Instance' dimension, is the process I call 'Growth of Consciousness'.

I believe that the process of growth of consciousness is hugely important - not only to us but to God, as well. If it is God's will and required goal that all experiences be experienced and that all things possible for energy to do are explored, then this process of growth of consciousness is one of the most vital aspects of the universe. It is 'the proof that a possibility has been explored' that is the proof that an experience has been experienced. It is the master checklist for determining God's progress towards Its goal of doing all that can be done with energy. In fact, growth of consciousness is the only other thing than entropy in the universe that continues to grow.

Perhaps it is growth of consciousness that is the offset for entropy. For every amount of energy that can no longer be used to 'work' in the 4-D space-time, an amount of 'growth of consciousness' is added, as well - simply because the work has been done, it's the natural and logical equal and opposite reaction to entropy. Thus growth of consciousness is extropic growth; the equal-and-opposite-reaction of entropic growth. Entropy does not seem to require consciousness directly but, in order to sustain any form of consciousness in our 4-D space-time, entropy will increase; therefore, consciousness does, indirectly, require entropy. Also, omniscience itself demands that God understands all aspects of Itself and that would naturally include the sum total of all energy that cannot be used for work.

It doesn't make sense, in my mind, for there to be only one thing that continually grows, as it seems to be in conflict with the laws of motion - specifically, equal and opposite reaction. But, in a system that obviously tends towards thermodynamic chaos and entropy, we can now fill in the previously missing opposing force: the tendency of the universe as a whole to create intelligent living entities that serve to increase the overall growth of consciousness. Only with growth of consciousness does the entire universal system finally make sense and our own existence as a form of life that is, for all intents and purposes, opposed to the tendency towards entropy, finally falls into place and is no longer seen as an aberration. This is not anthropic, per se, but it is intellithropic - a generally entropic system that is designed to counter that entropy by the production of a small number of relatively low-entropy entities, which are the various living things that are, at some level, aware of their environment.

As to how the energy in the Abstract Idea Space would look, visually, to human eyes if they were so placed as to be able to see it would be, I suspect a complex mishmash of incomprehensible brightness. It's raw energy used in a way that is not to be looked at but to be utilised by the mind. Its looks are irrelevant. It's the meaning behind the coding that is what is important and I don't think, for a minute, that human eyes would be able to see the coding; rather, I think they would be completely blinded by the brilliance of the raw energy and its constant motion from one state to another as the minds of conscious entities are constantly making changes to it. The coding itself is, probably, handled by topology that we may not have thought of yet. The way the mind utilises that data must be similar to that of relational algebra and calculus - similar to the forms used in our modern relational databases - that must be deeply encoded in the metadata of the mind of God and as a sub-process of our own minds.

To peer into the mind of God with simple human eyes would be a futile effort. It's not the kind of thing that eyes were designed to see. In much the same way, when one human looks at another, we don't see the individual cells of their body or the bacteria that clings to most of it or the emotional state the individual is in, for, if we did, we'd likely be put off by what, otherwise, would be an individual of outstanding beauty. The mind of God is, most likely, the most beautiful thing to behold; but, our eyes would never see it that way - it is only our minds that can reveal a small portion of that beauty. It is only with our minds that we have any chance to perceive or conceive of the true beauty and complexity of this Abstract Idea Space where the perfect and eternal forms of everything imaginable and real must first exist. Our other senses are simply not suitable in any respect for sensing the most sensible thing in existence - the mind of God.

How is God's Word Used to Create?

This is a theme that runs through the Jewish, Christian and Islamic scriptures, that is, that God creates by the power of 'God's Word'. When God speaks, matter moves and forms reality. The beginning sequence of creation in the book of Genesis is:

(Gen. 1:3) Ve-ayamar Elohim, "Yehi Aur" ve-yehi aur.

Which, when translated, means: And God said, "Be there light" and light was. The concept is that creation is achieved through the power of God's speech. Scientists usually scoff at this as being ridiculous and not even worthy to investigate, as it's based on some ancient anthropomorphic idea of God that God, like a man, can speak. While there is an element of anthropomorphism in the usual interpretation of this and similar passages, there needn't be out of necessity.

The non-anthropomorphic view is simple enough. Speech is, for all intents and purposes, sound waves - simple vibrations. The power of vibration is the underlying theme behind String Theory; so, immediately, it tallies with this model of physics. The strings vibrate at different rates and 'create' the appearance of various sub-atomic particles. These aggregate to form atoms and molecules on up the line. To my way of thinking, this isn't anthropomorphic at all; rather, it is a simple analogy that drives at the point that it is vibration that causes matter (and anti-matter).

The Semitic languages - the language used to write the Hebrew Scriptures, the language of Jesus' speech (Aramaic) and the language of the revelation of the Qur'an (Arabic) - are very unique in certain respects. The languages are based on the underlying verb stems. For the most part, each verb stem is

113

comprised of three letters, although there are a few verb stems that are comprised of four letters, the vast majority are comprised of only three. Now this bears a striking resemblance to the three-letter codons that are used by DNA to code for amino acids. Is it a coincidence or is it another reflection of our naturally holographic universe and that we create our proteins in a similar way to the way 'Holy Languages' are based on their verb stems? As you've probably guessed by now, I don't believe in coincidence and I use the concept of the space-time continuum to back that up. There are no missing points in the space-time continuum, so there can be no such thing as a coincidence, as all events are already planned and fully mapped out. So, what we are seeing are the parts of the plan, or, perhaps more correctly, the engineering features of the planning.

The Jewish Scriptures focus more on the power of speech whereas the Qur'an spoke first of the power of the written language by reference to 'the pen'. God's pen is, of course, unlike our pens in that God's pen writes on the scrolls of our souls. The very first revelation that came to the Prophet Mohammed (saw) were these lines:

Surah 96:1-5

96:1 Read! In the name of your Lord and Cherisher, who created
Iqra! Bismirabbika-lladhi khalaq

96:2 Created man from a clot of congealed blood.
Khalaq-al-insana min alaq

96:3 Read! And your Lord is Most Bountiful
Iqra! Wa Rabbuka-al-Akram

96:4 He who taught by the Pen
Alladhi allama bil-qalam

96:5 Taught man that which he knew not.
Allam al insana ma lam ya lam.

Whilst it is obvious that, in this case, God is speaking about the use of the pen to write scripture and how scripture can be used to teach men things that they would not understand otherwise; there are other analogies to the pen - written word - and spoken word that I will explain a little later. For now, I want you to consider that, with reference to God's 'power of speech' the letters God uses are the basic elements outlined in the Periodic Table of Elements. Is there a reasonable correlation between the 8 periods of the table and the 8 parts of speech? I believe there is; let me show you.

Let's look at language and see how it relates to matter. I think sentences act like molecules. Each one has a particular purpose, structure and quality. Yet they are made of words. That makes words akin to atoms. But atoms are further divided into the sub-atomic particles of hadrons (heavy sub-atomic particles) and leptons (light sub-atomic particles) that, like words, are comprised of letters which are either consonants or vowels. Yet even letters can be viewed as being made of lines, either straight or curved. Here, too, is a subtle, holographic allusion to String Theory and the concept of closed and open strings. Also, atoms (words) fall into 8 periods in the Periodic Table of Elements. These are, in a way, akin to the 8 parts of speech: nouns, verbs, pronouns, adjectives, adverbs, conjunctions, prepositions and interjections. Yet some elements fall into transitional groups. Theses would be akin to the concepts of participles and gerunds. A participle is a verb-like word that acts like an adjective, e.g., the word 'sinking' in the sentence: "Every time I see the film 'Titanic', I get a certain sinking feeling." The word 'sinking', although it is a verb, acts as an adjective to describe the word 'feeling'; and, because it acts as an adjective is, technically, a participle. The word 'feeling' in that sentence, although it is a verb, acts like a noun; and, because it acts as a noun, is, technically, a gerund. These are transitional parts of speech where one type of word acts as a different part of speech than it may at first appear and only context can determine the difference.

To see this more easily, I'll map out the parts of speech to the Periodic Table based on Semitic language. Firstly, it's easy to see that interjections (like 'Wow' and 'Hey') stand alone and do not combine with other parts of speech; therefore, the interjection is Period 8, the Inert or Noble gases. All Semitic languages have their root words as verbs. Verbs are conjugated, have tenses, number and person. They are the most configurable of all parts of speech and seem the most likely to sit at Period 1, as the Period 1 atoms combine with other atoms the most. Period 2, then, would seem to be nouns. In Semitic languages, nouns are formed from their root verb stems because every action implies an actor. The concept 'to build' implies the concept of 'builder' as much as the concept 'to speak' implies the concept of a 'speaker'. After Period 2 are the Transitional Elements. These are the verb forms that act as either nouns (gerunds) or adjectives (participles). Following that logic, at the other end of the Transitional Elements is Period 3, which must be the adjectives, because the Transitional Elements ought to sit between those parts of speech that they form their transitions. Now, let's go back to the other end of the table. Pronouns stand for specific nouns, that is, they each have a single antecedent, a noun upon which they depend and to which they refer and, thus form an obvious one-to-one relationship. This seems akin to the Period 7 Halogen group, as they can only combine with one other atom. Period 6 has two open places for connection with 'others' and so seems to fit in well with the concept of a conjunction, which links two 'other' things together. The Period 5 group has three open places for connection and seems a best fit for the concept of the preposition which can relate one object to another either directly or indirectly or both. That leaves Period 4 to pertain to adverbs. And, with that, each period is covered and directly corresponds to a part of speech. If you think I've left out the concept of the 'article', then think again. The Lanthanide group is most akin to the 'definite article', as they are all (well, with the single exception of Promethium) non-radioactive and are stable elements that do not degrade or change, thus, they are definite; I have no problem allowing God to have a definite article that is unlike human language by being greater than human language in some respect. This leaves the Actinide group to be representative of the 'indefinite article' as they are all radioactive and unstable and, in that respect, indefinite, precisely because they are unstable. And now, all parts of speech are covered by their corresponding aspect in the Periodic Table of Elements.

It is my hypothesis that God creates through these words or elements and it is on that basis that the concept of 'The Pen' relates to how God creates. This completes the examples of how God's creative Word can be analogous to fermions, that is, the sub-atomic particles like hadrons and leptons that comprise atoms/elements.

Yet there are subtle inferences that are implied. For example, the pen and the voice are the forces behind written and spoken language. And, of course, in each case, there must be an author and a speaker. These are the forces that act behind the pen and the voice. So, there are, behind this creative ability, four forces that are analogous to these: the four bosonic forces of electro-magnetism, gravity, and the weak and strong atomic forces. Of all of these, the analogy of 'The Pen' to the electro-magnetic force is the most obvious because a pen is useless without ink. So, as the ink goes with the pen, the electric and magnetic forces are always found together. The voice, then, must be most analogous to gravity, as it is unseen but moves us in ways unimaginable. This leaves the weak and strong atomic forces as being analogous to the author (weak) and the speaker (strong). I believe that the spoken word is stronger than the written word simply because one must learn to read in order for the written word to be understood,

117

whereas hearing is all that is required for the spoken word to be comprehended. Put another way, an illiterate individual can be moved by the spoken word but not by the written. Also, like the strong atomic force, the spoken word is only heard by those who can hear it (ignoring, of course, recordings OF spoken words, which, only in recent years, have made the spoken word reach farther); whereas the written word, like the weak atomic force, can stretch over longer distances across both space and time. Additionally, the written word can break down; books can decay and ink smears; and, so, we have another analogy that relates the written word to the weak atomic force as it is the weak atomic force that allows for radioactive decay.

So, how is God's word used to create? Ignore the anthropomorphic aspect and understand God's 'speech' as the vibrations of the elementary strings of String Theory. God uses those vibrations to put the underlying strings together to form atomic elements and molecules from them. God organises them into crystals or membranes or organs and some complete living organisms. This doesn't happen haphazardly throughout the universe; rather, it happens in places where the living organisms can relate to their environment in an intelligent fashion and, by doing so, continue living. Also, non-living things are not placed haphazardly but in the cases of galaxies and star systems, they are laid out in an extremely careful manner and abide by very strict physical rules. In fact, the living organisms must abide by the same strict physical laws. The combination of all the events in this universe is, indeed, incalculable by humans, but it is not random.

There is a great difference between randomness and incalculability; one doesn't need an intelligent force behind randomness but an intelligent force that expresses itself down to the quantum level will, invariably, produce incalculability. Equally, no human intelligence is powerful enough to tell the difference between whether or not we observe true randomness or simple incalculability; so, for that reason, we ought not to accept randomness in the face of all the order observed throughout the universe. Additionally, and far more importantly, a space-time continuum with no missing points cannot, logically, allow for any randomness because all the spatio-temporal events are already defined; so, randomness is, simply, something that does not exist in reality. The institution of Science uses randomness to explain away incalculability purely in order to save face. Let me state it again, randomness cannot occur in a universe governed by Special Relativity because a space-time continuum contains all spatio-temporal events; therefore, a scientist who accepts Special Relativity - and all of them should - is a hypocrite if he or she also speaks of finding randomness in this universe. I have no doubt whatsoever that the quantum state of the universe is incalculable. In fact, the quantum state of a grain of sand would be difficult enough to attempt to calculate even if you could completely isolate it from outside influences; of course, the measurements themselves would act as outside influences and alter the findings, thus ensuring a result that is imprecise. Yet the entire universe works like a well-oiled machine. However, there are still a couple of important facts that Science can't explain, like dark energy or the arrow of time. Given their inability in this regard, I'll proffer my explanations for these two important aspects of the universe in the following two chapters.

What is Dark Energy?

If I'm correct in what I believe dark energy to be, then it will be one of Science's greatest oversights and one of my greatest insights. I believe that the question of the nature of dark energy is related to two equations:

1) $E=MC^2$
2) $E=F\hbar$

The two equations above are: 1) Energy equals Mass times the Speed of Light squared and 2) Energy equals Frequency times the Planck Constant. Science is extremely happy with the veracity of these equations, yet it has somehow missed how the two work together to solve the dark energy problem.

Every star or every energy-producing sphere puts out energy in all directions, that's 360 degrees by 360 degrees by 360 degrees or, when speaking strictly about degrees, a star will radiate light in 46,656,000 different directions at once. This is 360 degrees across all three spatial dimensions. But, of course, the truth is greater than that. Every degree is 60 arc minutes and every arc minute is 60 arc seconds, so the total number of directions, in terms of arc seconds, in which a star emits photons, is a whopping $(360*60*60)^3$, which is 2,176,782,336,000,000,000 or 2.176782336 quintillion directions. But, of course, even arc seconds can be sub-divided and the real number of directions in which a star emits light vastly exceeds the quintillion figure above. For argument's sake, we'll stick to the quintillion figure, as its big enough to express my point.

We'll assume that every star emits streams of photons in that many directions every second of every day, yet only those pointed directly at us are 'seen' or 'observed' by us because we happen to be lined up correctly to see just those. All the other streams of photons pass obliquely across our path of vision and, because we are not lined up with them, we don't see them; they appear 'dark', thus the reason why we cannot see them. This, then, is the answer to the question, "What does a photon look like when seen from the side?" It doesn't look like anything, as it is completely dark; but that does NOT mean that it doesn't exist. The very fact that streams of photons pass by us obliquely on all sides is very important, for it is that which makes a shadow. Furthermore, these streams of photons have been doing this for millions if not billions of years - depending on whether or not my theory of temporal contraction is correct. Irrespective of that, though, they've been doing it for a very long time, indeed. And, we are not counting the fact that some of these streams of photons have been reflected or refracted by various things in various ways. As long as the photons have not been absorbed, they still play a role in this concept.

Each of these photons has a specific frequency. You take that frequency, multiply it by the Planck Constant and achieve the energy of the photon. You then use the other equation and divide that number by the speed of light squared and yield the mass of the photon. Now, we're taught that a photon has a rest mass of zero; but, when have you ever seen a photon at rest? If the photon is not moving, it can hardly be classified as a photon, as they travel as near to C as possible slowing down only due to the medium through which they are moving.

All this extra mass, the mass of all these streams of photons that pass obliquely past our line-of-sight, I believe, is what accounts for the 'dark energy'. Naturally, the more space through which you look, the more dark energy is measurable. And, of course, as more photons are produced by the stars in galaxies, you would, again quite naturally, expect to find more dark energy associated with galaxies, which is exactly what we observe.

Now, what I haven't done is to try to calculate how much mass has been created via these streams of photons since stars were formed. Rather, as a thinker, I've come up with the concept and I'll leave the 'busy work' to those who are better qualified to work out the numbers. Of course, this would only be making an estimation of the dark energy. In order to do this, we need to calculate, roughly, how many galaxies there are, how many stars per galaxy, how many stars emitting photons of varying frequencies and let's not forget the pulsars and quasars that emit huge amounts of energy. Also, if we are to attempt to calculate the total mass, we must not forget about the part of the universe that is beyond our observable universe. Just because it is beyond our observation does not mean it doesn't enter into the equation for the total mass of the entire universe!

If the mass that is calculated is off by even a factor as great as 100, I think we're on the right track, as that kind of variance could only prove that there are things like far-away brown dwarfs that are, to us, invisible, but they are still emitting streams of photons in all directions and add into the equation. Billions of quintillions of radio waves add up rather quickly. I really do believe that it's as simple as that. We've overlooked the obvious. We haven't accounted for the mass of photons that pass obliquely past us and THAT is the solution to dark energy. It isn't some unknown substance; it's the fact that there is a direct relationship between frequency and mass, ($ħ/C^2$), that has been known yet completely overlooked by Science for the past century.

Now, don't be angry at Science for overlooking the obvious; it's an occupational hazard of the practice. Science often misses the forest not for the trees, but for the chlorophyll in the leaves on the trees. Science is often carefully scrutinising the smallest of details and not standing back and seeing the big picture. Plus, most scientists are great at deductive reasoning but not that great at inductive reasoning and this answer calls for the latter. It requires lateral or 'outside-the-box' thinking, which is what I do or, it is, at least, what I try to do. I just hope that, when some scientist does take it upon him or herself to do the rough calculation I describe above, that the resultant figure will be close to the estimated amount of dark energy proposed by science to exist.

What's Behind the Arrow of Time?

As I've alluded to above, the arrow of time is not directed by entropy but, rather, by a more obvious and simpler aspect of the geometry of the system. The Standard Model predicts that magnetic monopoles should be a feature in our universe, yet no scientist has ever discovered one, except one unrepeatable instance. It's a bit like not being able to see the Higgs bosonic structure, as it 'exists' in a place we cannot see - around a dimensional corner. The Higgs structure is situated in the space that our space-time expands through and, if we could see it, we'd see little else, as the 'fabric of space-time', which is so finely structured and fills both the space-time we are in plus the space through which that expands, it would completely fill the space we look through and we would see nothing else but that structure. In the case of monopole magnets, we don't see them for similar but slightly different reasons.

Our entire space-time and the membrane that it forms is a magnetic monopole when looked at from the outside - that is, when looked at from within the space through which our space-time is expanding... if you could see in that space. The opposite monopole is the wall of anti-matter that forms the outer boundary of the space through which our space-time expands. The wall of anti-matter is fixed but the space-time is elastic. The difference in charge between the matter of our elastic space-time and the fixed anti-matter wall is what drives the expansion of space and, in turn, it is that expansion that causes time to flow forward.

It's really that simple. It's entirely possible that it would be better to describe our 'matter' as anti-matter and vice versa, as it is our negatively-charged elastic universe that is attracted to the positively-charged fixed wall. If the elasticity were the other way, the wall would collapse and rush towards the fixed blob of 'matter'; yet what we see is the expansion of space; so it is space-time that is elastic. This, of course, does not help us to determine which of matter and anti-matter is truly positively-charged and which is truly negatively-charged; the fact that they are of opposing charge and only one material is elastic is what determines the expansion of the elastic material. Whether or not it is matter that is North-seeking or anti-matter, either way, as long as our space-time is elastic and the wall of opposing material is of the opposite charge, the attraction will occur and the result will be the expansion of the elastic material and the flow of time perceivable within that time-containing material will move forward, thus, from the past towards the future.

So, rather than being pushed forward by the complex accumulation of entropy - a measure of the amount of unpredictability/disorder and useable energy in the system - the forward flow of time is a simple result of opposite charge electromagnetic attraction. This is a far simpler explanation and is, therefore, more likely via Occam's Razor. In fact, it even calls into question the concept of entropy. How do we measure when a quantum of energy can no longer be used for 'work'? Is it down to our ability to re-capture and store so-called waste energy? Would a more efficient solar battery, able to recover seemingly 'lost' and unusable energy, be a means of lessening universal entropy? If so, then entropy is more related to human technology than any objective physical process. Surely the passage of time is much more likely to be determined by two magnetic monopoles attracted to one another than by the amount of useable (by what or whom!) energy in the universe? Besides, energy considered as no longer useful is what combines with the background radiation to keep our inter-galactic space slightly above absolute zero; I would think that could be incredibly useful overall.

At a more microscopic level, the actual attraction is probably down to the fact that atoms have electron shells around them and these electrons are on the outer side of the atom. Electrons have a negative electrical charge. Their anti-matter opposites, positrons that have a positive electrical charge, are also in orbits around nuclei of anti-matter, i.e., on the outer side of anti-matter. The cloud of fixed positrons covering the wall of anti-matter is what is attracting the outer side of the atoms in our universe, i.e. the electrons, to it.

We exist on the very edge of the expansion. Put another way, if you viewed the expansion of space-time as an expanding balloon in typical 'brane-theory' style, we exist on the outer part of the membrane. Motion, as observed by us, only exists on this bow-wave of expansion. What lies internal to the expanding membrane is a smaller universe that represents our past; the future is what lies between us and the wall - we just have to expand through it to encounter it. Seemingly, the space we expand through is porous to space-time and the Calabi-Yau space, although the Calabi-Yau space serves only eternal functions and does not play a role in our perception of the flow to time. The anti-matter's opposite charge pulls the matter from the past and into the future through the grid-work of Higgs material that fills the space between the matter and the anti-matter. As time passes, our universe expands and we, riding that bow-wave of expansion, experience the flow of time rushing from the past into the future. Our relative future is locked in front of the bow-wave and we can't see it - but we can speculate about it. We cannot see the near past because we have expanded past it (no pun intended, but it helps demonstrate why we call it 'past') and the events are locked 'behind' the expansion; however, we have memory and can remember parts of that past and either speak of or write down what happened.

This explanation for the reason behind the flow of time and our perception of the flow of time is far simpler than any other explanation put forward and it is in complete agreement with scientific laws and probabilities, even though it is not in exact agreement with current beliefs about how the universe works.

Because it is simpler, I believe, it is more likely. It passes Occam's Razor and relies on basic mechanics. Only when you exist exactly on that ever-moving, bow-wave of expansion could you ever experience movement because that spatio-temporal membrane is the only moving thing with respect to the dimension through which it is expanding. The movements of matter within the membrane of our space-time continuum do not contribute to the arrow of time; rather, it is only the expansion OF that membrane that accounts for it.

As previously mentioned, perhaps it is the magnetic field force of the attraction between our space-time and the anti-matter wall that is the 'Cosmological Constant' that Einstein thought about for a moment and, then, discarded as irrelevant. I believe this may well be the case because, within the expanding space, that field of magnetic force is irrelevant because of Special Relativity; however, in the space through which our universe expands, it IS a functional necessity if one is to determine the life-time of any given universe and explain the entire system. Both the elastic space-time and the substance of the anti-matter wall are a part of the space-time continuum and, therefore, in order to understand the equations behind the continuum, one must take into account that the matter and anti-matter are both part of the space-time continuum, but, as only the matter-containing aspect expands, time is only perceived in it, as the wall of anti-matter is fixed and non-moving, even though it has a temporal aspect, which allows it to persist until the matter and antimatter re-engage in annihilation, no passage of time would ever be experienced within the anti-matter simply because it is fixed spatially.

I state that the wall of anti-matter is fixed in place but have not discussed why it is so. The answer to this involves discussing 'absolute nothing'. This may come as a surprise to many readers that 'absolute nothing' needs to be discussed; nevertheless, it is absolutely vital. What can be said about 'absolute nothing'? Well, rather more than you might imagine. In the cosmology I present, 'that which is' is surrounded by nothing. That is to say that there are two aspects to consider: 'that which is' and 'that which is not'. Our entire universe, incorporating all 26 dimensions, is surrounded by 'absolute nothing' and that 'absolute nothing', by its very nature, cannot be ignored, as it has an effect by not allowing anything to affect it. In order to remain 'absolute nothing', it cannot admit, permit or transmit anything or it would be more than 'absolute nothing'.

I ought to clarify what I mean by those three verbs and I think it is, perhaps, easier to see the subtle difference between them through their respective nouns: admission, permission and transmission. By admission, I mean acknowledgement, as in 'I admit that it's true'; by permission, I mean allowance, as in 'I permit you to do that' and by transmission, I mean conveyance, as in 'this wire transmits electricity'. The first two of these concepts refer to the fact that 'absolute nothing' has no consciousness and it has no consciousness in either an active way, through admission, or in a passive way, through permission; the lack of the ability to transmit is purely physical, as 'absolute nothing' is no form of medium and could never be used by energy. It should be clear that there is a huge difference between 'absolute nothing' and the 'vacuum space' that exists in our space-time continuum. If, for example, you were floating in inter-galactic space (vacuum space), you would still be able to see the light from the surrounding galaxies because 'vacuum space' transmits photons and, in fact, because of that, it is, perhaps, almost filled with photons. This is one of the facts that relates to dark energy and to the appearance of so-called 'quantum flux' in 'vacuum space'. However, true 'absolute nothing' is very much unlike 'vacuum space' because you could never get into it, as it cannot acknowledge your presence through admission, allow you to enter it by permission or allow you to pass through it via transmission.

So, in a very real way, 'absolute nothing' is that which is most repulsive to energy BECAUSE it cannot admit, permit or

129

transmit anything that is comprised of energy. This is why the anti-matter wall is fixed in place forming a barrier against that 'absolute nothing'. Matter and anti-matter react to one another through the process known as annihilation and both anti-matter and matter will react to its own kind according to the known laws of physics. But neither of these types of energy can, in any way whatsoever, react to 'absolute nothing'.

If, for example, a photon were to encounter 'absolute nothing', it would simply stop its forward motion because, as I've stated, logically, energy cannot react in any way to 'absolute nothing'; therefore, 'absolute nothing' acts as a barrier that is, in effect, more solid than any solid made of matter or anti-matter. So, in a similar way to how a photon would stop its forward motion if it encountered 'absolute nothing', so, too, would any form of anti-matter, as it is just another form of energy. Energy, which is 'that which is', cannot react to nothing nor can it penetrate it in any way. So, the anti-matter was separated from the matter and folded 'out' a fraction of a second after the Big-Bang and it was moved to the farthest extent of the space through which our space-time expands; it then stopped dead in its tracks when it encountered the 'absolute nothing' that surrounds 'that which exists'. That is why I say the wall of anti-matter is fixed in its place. It exists right up against that 'absolute nothing' and does not react to it; thus it forms a protective shell separating 'that which exists' from 'that which does not exist', which is 'absolute nothing'. I believe it is that fact that makes sense of the Native American belief that our universe is analogous to a tortoise. The anti-matter forms the shell of the tortoise and the matter is the living thing within the shell.

I believe that I may be one of the first people to ever spend six paragraphs discussing the virtues of 'absolute nothing'. In fact, I'm reasonably certain that I'm the first to discuss some of the vital properties of 'absolute nothing' and how they pertain to physics in a cosmological way. It is perfectly logical to conclude that 'absolute nothing', as I've stated, cannot admit, permit or transmit anything and continue to remain 'absolute nothing'. Also, in the minds of the readers, I feel I have to make clear the distinction between the reality of 'absolute nothing' from the concept of 'absolute nothingness'. In the Abstract Space, there, no doubt, exists a definition of the idea of 'absolute nothingness'; but, in reality, the real 'absolute nothing', which cannot contain anything, is, nevertheless, that which completely surrounds 'that which exists'.

I know it's a difficult thing to imagine, believe me. Nevertheless, it is vital to understanding how real 'absolute nothing' plays a role in this physics. Because neither matter or anti-matter can react with it, an all-wise Creator must prevent the possibility of matter encountering it; so, that all-wise Creator put a wall of anti-matter up against it so that the entire system of 'that which is' is, in effect, self-contained. The final reaction in a given 'Big Bang to Annihilation' cycle, then, is not that matter stops moving when it encounters 'absolute nothing' but is, rather, a matter/anti-matter annihilation that, due to the toroidal shape of the anti-matter, the continuing expansive motion forces the annihilating energy to follow the negatively curved shape of the torus and wrap back around to the Planck length-sized centre of that torus and begin a new 'Big-Bang to Annihilation' sequence; thus maintaining the continuation of Creation. Without both matter and anti-matter configured in this way, there would only be a matter-based, single Big Bang that expanded and stopped when it encountered 'absolute nothing' and that would not allow energy to explore all its possibilities. An all-wise Creator is smarter than that; and, I suppose, since I can see the logic too, so am I; but, I am so far from all-wise that, if I have a proton's worth of wisdom in comparison to the universe of God's wisdom, I should be so lucky. I also believe that most of the readers of this book will understand this logic too. Plus, it makes perfect sense of the philosophy of the

Yin and Yang opposing forces that are necessary for a logical whole, which is the Tao.

The wall of anti-matter, in all likelihood, is neither isostatic nor isotropic; rather, it is isomagnetic and isoelectric in relationship to the density of matter in the expanding space-time. So, the anti-matter wall, during the annihilation phase at the end of the universe, when the matter expands to the point of reaching the anti-matter wall, is thick enough in the right places to completely annihilate the matter. Areas of the anti-matter wall that will have to annihilate super-clusters of galaxies would have to be proportionally thicker in order to do that. Only by being distributed in an isomagnetic and isoelectric fashion could a perfect annihilation occur and that kind of distribution would make the entire matter/anti-matter system isodynamic with respect to their equivalent masses and their naturally opposing isomagnetic and isoelectric properties. That total annihilation coupled with the negative curvature of the outer boundary wall will return both substances together as pure EM radiation so that a new Big Bang could occur once the annihilated energy has moved around the toroidal shape of the anti-matter wall back to the central position that enables the entirety of energy to come squirting through the Planck-sized hole in the middle of the toroidal space and allow a new Big Bang to occur.

It seems to me that it should be possible to extrapolate the force of this magnetic field between the two opposing monopoles assuming equal mass in both. The closer we get to the wall, the greater the rate of expansion. You can see this is true by holding two magnets of opposite poles close to one another and moving them towards one another; there comes a point at which the magnetic force increases and is so overwhelming that it pulls the two completely together. Whilst it may not be increasing quickly, it does appear that the rate of spatial expansion as dictated by the Hubble Constant, which we know to be a variable, is increasing. This would be the case if our space-time were nearing that wall of anti-matter. Perhaps by calculating the rough strength of the magnetic field that is caused by the matter/anti-matter attraction, it would be possible to calculate the distance to the wall and, thus determine how long our matter-based space-time has left.

Whilst I could calculate this value, I would not publish it as I believe any 'end time' dates given would only cause panic. From a philosophical standpoint, I believe we should act as though every day were our last and try to live each day the best way possible. I see no advantage in telling the populace when they're all going to die, as it would cause many to act in ways they otherwise wouldn't and I do not want that responsibility on my head. So, if any scientist wants to calculate this based on my hypothesis, feel free; but be careful about publishing it because, if you (and I) are right, then you may set off a ripple-effect of events in people's lives based on the panic ensued and, as I said, I want no part of that. Yet I can't inform the world about the mechanism behind the arrow of time without also implying that, whilst the space-time continuum will exist for all time, each individual Big Bang-to-Annihilation sequence will have a beginning and an ending.

Everybody knows that, at some point, they will die; but, people are used to not knowing when or how. To inform people of the when will change the perspectives of many people just as knowing how but not when one will die might change a person's lifestyle. But, in this case, there is no way to alter the fact of matter/anti-matter annihilation if I'm correct about the geometrical configuration of the universe. When this particular cycle reaches its end-point should be, philosophically, a fact that would not change the way wise humans live their lives. But we do not live in a world where most people are that wise or emotionally intelligent. If you care about the world and you still want to know the end-date, calculate it for yourself; just don't publish it. If you do publish it; it will have its effects and it will not be because you weren't warned.

Did God Create the Universe in Six Days?

This, I'm sure, is an argument that Science is very sure it could never lose; however, it is only due to a fair bit of Wizard of Oz-like "pay no attention to that man behind the curtain" that allows them to get away with it. What they will do is to draw your attention to the fact of radio-carbon decay and/or simple radioactive decay as the 'proof' for the age of the universe or the age of Earth. But this is only part of the equation and a part that becomes completely irrelevant when the full body of physical laws, as we know them, is taken into account. Relying on radioactive decay is a scientific hand-waving gesture that draws our attention away from that which really does matter.

The relevant part of physics - the set of equations that form our understanding of scientific truths - to which Science will not draw your attention is that of the varying rate of the expansion of space-time and this is the crux of the matter, namely, the variable that is called the Hubble Constant. In fact, since they discovered it varied, they have not bothered to rename it AS a variable but still refer to it as a constant when we really know better. To be fair, Science itself may well not have even realised the relevance the Hubble Constant has in this regard. In fact, I have met no one, to date, that has realised the importance of the Hubble Constant with respect to the age of the universe or its inherent relationship to radioactive decay. It's another case of Science overlooking that which has been staring them in the face and, once again, it is my heartfelt duty to point it out.

In 1929, Edwin Hubble discovered what we now call the Hubble Constant, the value that represents the rate of expansion of space in our space-time. This allowed us to determine the age of the universe based on the distance to far-away objects accounting for their red-shift via the Doppler Effect. However, as the Standard Model took shape over the next few years, it was discovered that, shortly after the Big Bang, the universe underwent a period of vast expansion, now commonly referred to as 'The Inflationary Period'. What this meant, though, was that, in order for space-time to expand at a vastly faster rate, the Hubble Constant couldn't actually be a constant; rather, it was a variable. As I mentioned in the previous chapter, today's measurement of the Hubble Constant shows that, even since 1929, the universe seems to be expanding at a slightly faster rate than it did when Hubble first discovered the value. This last fact is vitally important in my overall theory, as it has certain implications regarding Al-Qiyama (the Last Day - the day that the matter/anti-matter annihilation occurs), but, for this argument, those factors are irrelevant.

The fact that the rate at which space-time expands is variable rather than constant is the crux of the argument. We have no idea of the value of the Hubble Constant between the Inflationary Period and the time when Hubble first measured it in 1929. And this covers MOST of time. Once the Inflationary Period was over and the anti-matter removed from the system, the laws of normal physics would preside and events would occur as they did, irrespective of the value of the Hubble Constant. And both of those events took place well within the first second after the Big Bang. There is no evidence, nor can there be gathered any evidence, that would indicate the value in that meantime due to the effects of Special Relativity within our space-time.

There is, simply, no way of knowing or inferring the value of the Hubble Constant between the Inflationary Period and Hubble's first measurement, as all radio-carbon or radioactive decay is directly relative TO the value of the Hubble Constant, which governs the rate of spatial expansion at the time of the measurement OF the radioactive decay; and, as time is not separable from space, it affects the rate of the passage of time. All evidence that relies on radioactive decay only shows that time is relative and that is a fact of Special Relativity, i.e. both space and time are relative to the rate of C, the speed of light in a vacuum. Radioactive decay, whether it is radio-carbon decay or otherwise, offers us no clue as to the actual age of the universe; rather, it only provides evidence of relative age based on the current value of the Hubble Constant.

Let's use an analogy of space-time as a film, where we and everything that exists in the universe are actors in a vast film that runs from beginning to end and we are, by general accounts, fairly far into the film. There is no valid, scientific reason that the film couldn't have been on what would be, relative to the current value of the Hubble Constant, 'fast-forward' between the Inflationary Period and about 10,000 years ago. In truth, until Hubble measured it, it could have been at any value from slightly above 0 to what it was during the Inflationary period. The value of the Hubble Constant during the inflationary period was, exponentially, 129 times greater than it is now. That is, a '1 with 129 zeroes after it' times larger. And Science has no problem with that. The Hubble Constant must have had that large a value for the inflationary period to have happened and the world of Science is satisfied with that. But, what value does it take to cause most of 13.5 billion years to seem to pass in 6 days? It takes a value of the Hubble Constant to be about '1 with 31 zeroes after it' times greater than it currently is; which is immensely slower than what is required to explain the inflationary period. Yet Science doesn't seem able to accept that Creation could have happened in 6 of our day's time when the actual requirement is between the huge number required for the Inflationary period and the current value.

Common sense would dictate that the value of the Hubble Constant must have slowed over that period, but the rate of slowing is beyond our ability to observe, calculate or infer. So, what's so difficult to believe about the concept of a film moving in fast-forward? When it comes to the numbers game, it's far easier to accept a smaller number than a larger number, so 6 days is, in a numbers-game style, more likely than an inflationary period; yet, we know with a fair amount of likelihood, that an inflationary period existed. Whilst, mathematically, the odds that the rate of expansion must have decelerated are 1 to 1, we have no idea of the rate at which the slowing occurred and no radioactive decay would ever be indicative, as any rate of radioactive decay is directly proportional to the value of the Hubble Constant at the time of decay, which will be relative due to Special Relativity.

The slowing could have been smooth or rough. Unfortunately, we are not in a position to be able to retrieve past values of the Hubble Constant and there is absolutely no way to infer the values because of Special Relativity. We can only assume that space has always been expanding at some value and there's every reason to believe that. It would be a natural outcome of having an elastic space-time that is porous to the dimension through which it is expanding and that the matter within space-time is being dragged from the past into the future by the gentle but persistent tug of the opposite charge of the fixed, anti-matter wall at the edge of 'the larger universe'.

There is no and there can be no evidence found that could prove that a 6-day Creation is not possible because the value of the Hubble Constant must have exceeded the value that it would take to create the universe in 6 days during the inflationary period by a exponential factor of nearly 98. Just like the actors on a film don't know when they are on fast-forward, so, too, the dinosaurs would not have known they were on fast-forward. Nor do I suspect it was foremost in the minds of cave-dwelling hominids. Even the greatest of past astronomers would only have noticed changes in the universe that were and are accountable by Special Relativity. To them as to us, time is relative and will pass at a rate that would be comprehended as 'normal' to any individual at any time irrespective of the actual rate of the passage of time. This is, again, another one of the beautiful resultant effects of Einstein's Special Relativity.

Whether or not any 'temporal contraction' is occurring at the same 'time' as spatial expansion is unaddressed by the Hubble Constant; The Hubble Constant itself speaks only of spatial expansion and does not consider that time is inseparable from space. This, too, I believe is another oversight as space and time cannot be viewed separately. Either there is temporal contraction at a rate roughly that of the inverse of the Hubble Constant or pure, spatial expansion will demand that photons would be required to accelerate in order to cover the extra distance incurred by spatial expansion and still seem to move at an apparently constant value. In other words, either there is temporal contraction or photons must accelerate in order to appear to move at a constant rate due to Special Relativity.

If there is temporal contraction, then the universe is a lot younger than it appears to be. And, as I stated above, there is no way of measuring the variable rate of the Hubble Constant between the Inflationary Period and 1929. Hence, there is no reason to think that the time from Big Bang to mankind roughly 10 to 12 thousand years ago wasn't 6 days of 'our time', as the entire universe could have been on a relative 'fast-forward' and there is no way to determine the exact value of the Hubble Constant in the past at any time after the Inflationary Period. In fact, the value it had during the Inflationary Period was calculated by inference based on what MUST have been the case in order to give the 'cosmic soup' enough room to allow the laws of physics to take over from there. So, whilst the value can be inferred within a certain degree of tolerance during the Inflationary Period, there is no way to infer its value after that point up to the point of Hubble's own discovery. Because the entire inflationary period lasted a very small fraction of a second, this leaves it well within the realms of possibility that it was, indeed, a period of 6 'current 24-hour day' days that was in-between the Big Bang and, say, 10,000 years ago.

And, whilst some scientists might find it hard to stomach, they could never produce any actual proof against it.

Because the Hubble Constant isn't a constant, it allows for the possibility of a 6-day creation and, therefore, modern Science, taking everything into account that is relevant, cannot safely make the assumption that a 6-day Creation is simply impossible. Both the Torah and the Qur'an state that God says this is what happened. Now, that is not a scientific truth it is a religious truth. It serves as the only evidence to the fact. Science cannot accept it - not because it is unfathomable to the conscience - but primarily due to the fact that it can't be reproduced and it is, therefore, not scientific evidence. It can't be reproduced because there is an historical component to the event itself. You can't always have Big Bang creations happening at seemingly random times and places because it requires all the energy that exists to produce a proper one. Mini-Big Bangs are simply a funny thought experiment more ghostly than that of 'Schrödinger's Cat'.

The fact is that Science believes a 6-Day Creation impossible and completely disproven that that period could have been 6 of our 24-hour days; however, with only the usage of what modern science accepts as true and a broader understanding OF the implications than those scientists themselves seem to have, I believe I have proven that conclusion to be faulty. The misdirection of not mentioning the importance of the Hubble Constant, which is, in fact, known to be a variable, the hand-waving man-of-Science, like a stage magician, can point at radioactive decay as proof, when, in fact, radioactive decay is completely relative TO the value of the Hubble Constant and, therefore, stands as completely irrelevant to the argument. Thus, the proof Science values so much as an argument against a '6-Day Creation' isn't even germane to the argument; however, it takes a broader understanding of the philosophical implications of physics and the problem as a whole in order to see that.

I have not, of course, proven that it was 6 days; however, I have demonstrated that it is well within the parameters of accepted science that, mathematically, it is possible - but that does not make it so. The evidence for it is purely of a revelatory nature and, given that revelation is possible given my model of physics, I believe we ought to take that evidence into account. Still, this firmly removes any claim to evidence against a 6-Day Creation via radioactive decay and that is very important. There is absolutely no reasonable evidence that points away from the possibility of a 6-Day Creation once radioactive decay is shown to no longer be germane. Of course, there is scriptural evidence in favour of it; but scriptural evidence is not scientific evidence. Yet, as I pointed out at the beginning of the book, they are different kinds of truth. So we have one type of evidence for a 6-day Creation and no germane, scientific evidence against it.

In effect, it would make no real difference to anyone's lives if there were a 6-Day creation that ended some 10,000 years ago or whether it took 13.5 billion years to reach the day; the day itself still has to be lived. I have demonstrated that Science has no real hard evidence against it; rather, Science just feels really uncomfortable with a 6-Day Creation. But knowing that the rate of the Hubble Constant varies and that we don't know that rate of slowing between the Inflationary Period and 1929, we have to admit 'scientific defeat' when arguing scientifically against a 6-Day Creation. For those who are less than omniscient (and that covers all humanity), it is better they accept that they will never know everything. Even I find the anthropocentric aspect of a 6-Day Creation to be uncomfortable but not if the motive was to allow intelligent entities to recognise the existence of God; so, I temper that discomfort with the thought that there are, most likely, other intelligent entities other than humans in this vast universe and the 6-Day Creation becomes intellicentric rather than anthropocentric and I can live with an intellicentric universe for entropic reasons even before spiritual reasons.

PART 2
Philosophy

How Can Something Appear From Nothing?

This involves the age-old argument referred to as 'Creation ex nihilo'. There is no logic behind something coming from nothing; but, that does not preclude that something could come from the appearance of nothing - the difference is strictly in appearance and the answer to the age-old question could just be an amount of geometrical illusion on the part of the Creator. I derive my view of creation ex nihilo from hints in Genesis 1:1 and the Gospel of St. John 1:1, as well as my own understanding of my physical model.

Genesis 1:1
In the beginning, God created the heavens and the Earth.

Gospel of St. John 1:1
In the beginning was the Word; and the Word was with God; and the Word was God.

Genesis 1:1 is simply stating, grossly, that 'it all started', just take it for granted (although, if analysed kabbalistically from the original Hebrew, it will reveal many more details, but that's outside the scope of this particular exegesis). The Gospel of St. John 1:1, though, takes us through a very Gnostic and logical means of deducing the process of creation from nothing, to which I'll return near the end.

I'm going to present this argument as if it were the philosopher Baruch Spinoza presenting it, albeit with full knowledge of our current scientific understanding and full knowledge of my theory as presented above; in other words, I will employ a geometrical progression similar to Spinoza's *Ethics*, as it seems the most sensible and sane way to present this material because we need some definitions from which to make some assumptions (axioms) and derivations (postulates). Some of these postulates could be considered very Platonistic, but, nonetheless they are completely rational and logical. Plato was a very clever man and, in my opinion, about 2,300 years before his time. So, now I hand things over to Spinoza and we'll begin the geometrical progression; however, the parentheticals are purely mine.

Definition 1) By a 'thing' I mean that which can be expressed by a noun, that is, a person, place, thing (in its usual sense as opposed to the other three) or idea. Things that are ideas - abstracts, if you will - do not require a physical, tangible presence in order to exist whereas the other three types are all tangibles - concretes, if you will. Some may refute this and think that an idea requires a conceiver, I intend to demonstrate that only 'the possibility of its conception', an abstract concept in itself, is required in order for it to logically exist as a potential. Genesis 1:1 provides a hint (a 'remez' for those kabbalistically attuned or those familiar with Jewish exegesis) at this through its reference to that which was created being 'the heavens and Earth': 'the heavens' representing abstract existence and 'Earth' representing concrete existence.

Definition 2) By 'energy' I mean any of its various forms from potential extending throughout all forms of kinetic energy - including the gravitational, weak and strong atomic and electro-magnetic forces - and all the transformations and means of transference of energy from one form to another.

Axiom 1) Energy exists.

Explanation 1) The refutation of the existence of energy is patently absurd in a blatantly Cartesian manner. To deny that energy exists while content to purport one's own existence would imply that the speaker must believe that they consist of something other than energy and nothing other than energy has ever been discovered to exist, thus the blatant absurdity of the denial of the existence of energy by a 'thinking thing' given today's understanding of the universe. All things consist either as some form of energy or of some combined form(s) of energy. Most concrete things are comprised of very complex forms of varying organisation in a state of constant flux with their equally concrete environment.

Axiom 2) Energy can exist in a state of potential.

Explanation 2) I mean this as we normally consider potential energy in modern parlance. That is, for example, the kind of energy that is added to an object when lifted off the ground. The object now has the gravitational potential to return to the ground at, potentially, varying speeds, depending on any outside forces impinging thereon, but especially gravity in this example; however, one could, for instance, throw the object back to the ground and that would add additional non-gravitational energy. Potential energy has no 'appearance', that is, it cannot be perceived visually. The object lifted from the ground looks the same as it did when it was on the ground, but it has gained gravitational potential energy by being lifted above a centre of gravity, if not then thrown or otherwise somehow projected adding even more energy. So, potential energy doesn't look like anything but that doesn't mean it's not there. (Kabbalistically, this is referred to through the property of air, because it surely exists, as it can be felt as wind when forced, but cannot be seen. Thus, in Hebrew, the terms for breath, which is air breathed through the lungs, and spirit are the same, RVCh or Ruach, as it is felt that one's Ruach, one's sense of self, surely exists but cannot be seen. In this respect, the Jewish faith has been Cartesian in thought millennia before Descartes!)

Axiom 3) Energy can exist kinetically.

Explanation 3) I mean this as we normally think of energy in a state of motion. For example, photons emitted by stars that travel through space and enter our atmosphere and reflect and refract and are finally absorbed, perhaps by one of our own retinae, as we view the distant star. Or, to give another example, the Earth itself is kinetically rotating on its own axis while revolving around the Sun and, with the Sun, moves around the galaxy and the entire galaxy is moving through space-time.

Axiom 4) Energy can exist in a number of transformation and transference states.

Explanation 4) By this I intend all the various ways energy can transform from one form to another. Heat, for example - a form of electro-magnetic energy in the infra-red wavelength range - is transferred by conduction, radiation and/or convection. In any of these cases, electro-magnetic energy is moved (kineticised, if you will) from one area to another by transferring through a medium in one of these ways. Air, can be moved by sound waves, so, yes, the tree that falls in the woods with no one to hear it, does, indeed, make the same sound it would irrespective of the presence of an audience. At a more atomic/molecular level, energy can be moved by the passing of electrons or, in the process referred to as 'annihilation', energy that is particulate of opposite charge, e.g. an electron and a positron, is transferred into varying wavelengths of electro-magnetic energy, e.g. gamma rays, x-rays, light, heat, microwaves and radio waves but also sound, if there's a sufficient medium for that (space-time itself is the medium for electro-magnetic energy, which I believe is interesting to note).

Axiom 5) Energy can neither be created nor destroyed; rather, it can only be transferred or transformed.

Explanation 5) This has been a cornerstone of modern science - the conservation of energy - and has never, to date, been shown to be false. And, with the advent of string theory, if it is ever demonstrated that energy can be 'lost' from our 4-D space-time, that's no 'proof' that it has not been conserved in one or more of the other dimensions afforded by the theory. Equally it could be said that if it is ever demonstrated that energy can be 'found', that's no proof that it wasn't 'borrowed' or 'escaped' from one or more of the other dimensions afforded by the theory. A measurable gain or loss of energy would, however, be an indication that energy may be able to exist elsewhere, other than our 4-D space-time, which would lend credence to string theory as an explanation FOR that gain or loss, as no other theory has any other places or dimensions from which energy can come or to which energy can go. Exactly what energy can do or does do when it is in those other dimensions is out of scope here. (Yet, it is in my overall theory that it is in the 'non-4-D, space-time' dimensions of the Calabi-Yau space that currently contain - and have always contained and always will contain simply because energy is neither created nor destroyed and is, therefore, eternal - the 'timeless' abstracts and areas of consciousness.)

Postulate 1) The universe, as we know it today and including all that has come before, must have been, at some point, potentially possible.

Scholion 1) To think the universe, as it exists now, was never potentially possible is ridiculous, if not simply by A1-A5. Rather, it must have always been a potential possibility. So, we can safely assume that the entire actual (and by actual, I mean the real history, complete with all the concretes in existence up to this point in time, not history as we may have heard or discovered about it, which may or may not be entirely accurate) history of the universe has always been potentially possible. And, given that the state has occurred, the universe, as it exists now, was not only possible but also very likely.

Postulate 2) All abstract things that can be real are always potential.

Scholion 2) Since it is obvious that one can imagine a container without one actually being present, it stands to reason that the concept of a container does not require concrete existence; rather, it only requires the possibility that it can be conceived, which can be derived from P1, as a subset. Because we, ourselves, can conceive of a container, we must concede its potential existence. Time is no barrier to this, simply because cavemen didn't drive Formula1 racing cars does not mean that rubber wasn't always a possibility. Even before rubber trees existed, they, too, must have been possible and their very existence proves it.

Corollary 1) As a subset of the things that currently exist, all abstracts, as they require no concrete existence, can exert their influence whilst only potential.

Example: This can be demonstrated by the countless experiences of people having dreams of things and, then, inventing them; for example, many of Stephen King's novels and subsequent films were inspired by dreams and they have, in turn, inspired some of us to spend time and/or money reading or watching them. Alternatively, concepts that 'arise' or 'occur' to us that, then, allow us to understand something new, like the kind of intellectual epiphanies of Newton and Einstein also stand as examples of this. These abstracts had no physical form, yet they have had profound influence on, at least, our planet. So, too, there is no evidence that this influence would be absent in energy that had the potential for it, as well as the potential for its conception, which are both derivable subsets of P1.

Postulate 3) All concretes require their corresponding abstracts to also exist.

Scholion 3) For example, to think that an individual can exist without the concept of that individual also existing is unreasonable, if for no other reasoning but Cartesian reasoning. Similarly, the concrete existence of a bottle of beer implies the potential existence of the concepts of beer and bottle, which, in turn, are forms of the categories of liquid and container, as well as they are representative of all their component ingredients and skill in composition and the potentials involved in them. Of course, if the basic forms of energy to make beer or bottles did not exist, e.g. silica as the basis for the glass of the bottle, as was the case some 0.1 seconds after the Big Bang, no actual bottles of beer could exist, but the potential for their existence existed, as can be derived from P1.

Corollary 2) Actual, concrete awareness would only be possible if it were, first, an abstract possibility.

Example: By P1, the particular instance of consciousness/awareness we know currently exists and we can derive that, by P3, our consciousness/awareness is derived as a subset of the potential for it to exist by P2.

Postulate 4) Logic, as a very small subset of 'the things that are possible', was, at the same point as P1 via C1, potentially possible and exerted its influence on the energy that exists.

Scholion 4) Of all the subsets of abstracts that are possible, logic and the concepts which precede it, for example, categories, are the most important abstracts in existence, as they govern the rules with which energy must comply. Even the constants that govern physics comply with logic, which is why we can rely upon them. Logic rules and it always has, since there has always been the potential for it.

Postulate 5) All energy was, at some spatio-temporal point, completely potential.

Scholion 5) From A5 and P1, we can derive that, given all of space and time, there would have been a spatio-temporal point at which all energy would exist in a state of complete potentiality, simply because the possibility of that existed.

Postulate 6) Logic exerted its effect at the point of P5.

Scholion 6) Logic, as it requires only potential energy to exert its influence, by C1, did so. There were no concretes to prevent it from exerting its influence and no other abstracts do negate the effects of logic, so it can be assumed that energy conformed to logic even when all energy is in a completely potential form.

Postulate 7) The logical property of negation transformed the energy at P5 from being completely potential to begin to actuate, or kineticise, and realise that potential via the means afforded as a subset of A4.

Scholion 7) The potential energy, which contained the complete potential for all that energy can do was negated and, it was no longer potential. From A5, the energy that has always existed must have been forced from its purely potential form at P5 to an actuated form, this P7. This is the essence of creation-ex-nihilo. At this point, even the self-awareness of the existing energy, by C2, began simply because, prior to that point, it was possible and potential, and, thus, led to the potential for all concrete awareness that has derived from it.

That concludes the geometrical progression that leads to an explanation for creation ex nihilo and, so, I return to my own style to continue to elaborate. From A5, we know that energy always exists and we know, from C1 that abstracts require only potential energy in order to be effective. Through the power of logic alone, in particular, negation, at point P5, would force the energy to move from a potential form to a kinetic form, P7, via A4, and, thus, a Big Bang, which would be the very first Big Bang in the sequence of Big Bang to Annihilation sequences I purport in my model, would appear as energy 'apparently' sprung from nowhere, as potential energy has no 'appearance'. This energy expanded to produce our current universe according to the other abstract laws of physics currently defined by several constants all of which must have been possible and, indeed, be exceedingly likely. Surely, we must acknowledge that we are possible and, considering our existence, fairly likely.

The Gnostic, St. John approach is a lesson in practical mathematical ontology. Yuck! What I mean is that it boils it down to a logical equation. "The Gospel of St. John 1:1" hints at the concept of creation-ex-nihilo in a particularly Gnostic and mathematical way. If, for example, the 'Word' was 'not' in its meaning of 'that which is not', and, at the point of P5, the corollary, "all possible concretes do not exist but are potential", were both in effect, we can simplify these two concepts and put them together by representing 'the existence of nothing concrete' by the word 'not', as in "'that which was' was 'not'". So, The Gospel of St. John 1:1 says, "In the beginning was the Word"; that particular word is the word 'not' in its sense of representing 'the potential for all things, yet none of which are concrete'.

But 'not' is also the logical verb of negation and, when applied to the previous 'not' - representing that potential existence - gives us the next phase of the operation, "and the Word was with God"; thus, the mathematical expression "not(not)", applying the negation process to the potential energy and creating all that is possible (i.e., "not(not) = all things possible"). This lends credence to the fact stated by the third phrase, "and the Word was God." The Word, in one sense, became God "The Creator" by virtue of it creating, i.e., actuating the potential, the concretes that exist in our 4-D space-time. And this particular universe is, most likely, just a single universe with potentially innumerable prior universes and innumerable future universes.

One object of the above geometrical progression was to demonstrate my point of view that logic applies and constrains God, as God is (he lightly says) nothing but energy; although that energy is expressed in all ways possible and 'that which is possible' is governed by certain laws and logic. I feel that resolution of creation-ex-nihilo demonstrated the way in which logic constrains energy reasonably. Whether or not it satisfies anyone else's criteria is only a reflection of the beauty of the variance of God's many views and certainly doesn't negate the concepts above, but the above geometrical progression should, at least, serve to present my own resolution and a rational, working resolution. Naturally, I suspect criticisms of all varieties and they are welcomed if for no other reason than that I fully recognise that part of being thought provoking is being provoking; nevertheless, I would think it hard to find the progression above to be an impossible line of reasoning.

Is the Universe Teleological?

This chapter should be one of the shortest, if not the shortest of the book; however, I do tend to elaborate, so it isn't. The reason that it should be short is that the argument relies on already empirically proven evidence, although I will admit to there being one potentially arguable point. To those who don't spot it below, let me be intellectually as honest as I can be and point it out now: it lies in my belief that energy cannot support an infinite series of spatio-temporal events. I believe that there is, in fact, a limit as to what energy can do and I base it on just a few logical examples. For example, energy cannot be configured in such a way as to create a 3-dimensional, spherical cube that an individual can hold in their hands. Equally, energy cannot be configured in such a way as to allow for a human to be chromosomally both XX and XY in each non-germ cell. It can allow for someone to be XXY and, perchance, even XXXY, but that is not the same as being both and only XX and XY at the same time! If you find an individual that has two nucleoli in each cell and one is XX and the other is XY, you're patient isn't human and, most likely, is a real male-to-female/female-to-male shapeshifter of a gender kind. Move away from them quickly and call for backup. There are countless other such logical absurdities that energy cannot be configured to permit and it is with firm reliance on these restraints that I believe that there is a limit as to what energy can do and I believe that that is a reasonable assertion. So it may 'appear' to stand as a logical hole; however, I believe the likelihood of energy being limitless in its capabilities is 1:1 against, for the simple fact that I, a mere human, can conceive of at least one thing that energy cannot do and that proves a limitation; therefore, the capabilities cannot be limitless. Whilst I admit that the capabilities of energy are extremely vast to the point of being countless, there are known limitations and, so, I stand on the side of limitation for the above logical reasoning. Now, let's get on with the argument.

Whether or not the universe is teleological has been a great philosophical debate over the centuries. Firstly, let me explain, for those who do not know, what a teleological universe is. It is a universe that leads to a predefined end point and a universe in which all events before the end are designed to lead up to that end. In other words, it's another angle on a designed universe because the 'end' is a known goal, albeit a goal only known to the Designer. In this case, a universe that, inevitably, ends with a desired result. But, of course, this begs the question, 'Desired by what or whom?' Well, "the Creator of the universe" is the only logical answer. Science has fought against this concept for years simply because it invites - nay, invokes - a Creator into parley. And, like Voldemort, God is to scientists, He-Who-Must-Not-Be Named and, if God IS mentioned, scientists twitch, quiver, quake and shriek just like the wizarding population of the *Harry Potter* world. Certainly, if God cannot be the right-hand side of the equation, the answer, God cannot exist in any way on the left-hand side either, as that implies mentioning the name. And, as Ms. Rowling so correctly stated in those novels, that "Fear of the name increases fear of the thing itself", it would seem that scientists are the most God-fearing people on the planet, but for all the wrong reasons.

The time has come, though, to lift the veil that the world of Science would like to keep over your eyes. Einstein's Theory of Special Relativity describes a universe in which space and time are joined together into one, 4-dimensional whole called the space-time continuum. This space-time continuum contains all of space and time and Special Relativity has been proven empirically via time dilation experiments that return results predicted by the

Theory of Special Relativity. So, the Theory of Special Relativity has been empirically proven and, with it, that the universe, being a space-time continuum, contains every spatio-temporal event from beginning to end.

The philosophical implication of this is that the universe MUST be teleological if both the beginning and end points are fully defined within the whole. Therefore, the universe is, most certainly, teleological. We've known this since 1905 and I'll brook no denial of it. So, I'm afraid it is a case of Science accepting this philosophically as well as acknowledging it as factual. A scientist who accepts Special Relativity must conclude himself or herself as a believer in a teleological universe.

Even if we expand the space-time continuum to include repetitions of serial Big Bang-to-Annihilation cycles, the result is the same, because the space-time continuum contains ALL the spatio-temporal events included in the entire series of cycles from beginning to end. So, even at that larger level, the universe is STILL teleological, as the ultimate beginning and end points are defined as well as all the intermediate beginning and end points for each individual cycle. The argument was over in 1905; this is a teleological universe.

Science itself, via Einstein's Special Relativity, has proven the universe to be teleological; yet the world's greatest scientists seem unready or incapable of accepting the natural, philosophical implication, a teleological universe, because it tends towards admitting a design thus a Creator; so, their stiff-necked opposition to the concept is in complete disagreement with known fact. And, as I feel it's my duty to blow the lid on these kinds of cover-ups, whether they are done knowingly or unknowingly, I now feel vindicated with respect to the concept of a teleological universe. I say unknowingly because I suspect that there are a great number of scientists that simply have not considered the philosophical implications of Special Relativity; although I myself saw the implication immediately. Personally, I feel a duty to inform the public at large because these implications are, in some cases, huge and change the entire way we must view the universe and our role in it.

The first thing that goes in a teleological universe is free will as generally perceived and this affirms the concept that Special Relativity also solves the question of Free Will (although I will address this topic in-depth in a later chapter). This, I'm happy to state, is also completely internally consistent and internal consistency is precisely what you would expect out of a correct and truthful view of reality, which I purport that my model is. How can an omnipotent entity allow any of Its creations choice, as choice implies a power to choose? An omnipotent entity cannot, logically, give up any of its power and remain all-powerful.

But we are still left with a question. Is there an infinite amount of time? If so, then the continuum, like the continuum of whole numbers, becomes an infinite series and no end-point can be defined. However, I believe that this can be resolved by a reasonable hypothesis. The crux is the answer to the question: What is it possible for energy to do? Now, no one will ever have the time, in one life-time, to divulge the true answer to that question; however, I don't believe that energy can do an infinite amount of things. That is, I believe it is folly to believe that an infinite array of spatio-temporal events can be affected by energy. Rather, there is a limit as to what energy can do. And, if there is a limit to what energy can do, the question of there being an infinite amount of time no longer becomes germane to the argument.

Once we have reached the limit of what energy can do, irrespective of how much time there is, then all the possibilities have been exhausted and the only choice left for 'The Creator' is whether or not to do them again. It seems ridiculous to me that God should be forced to perform everything an infinite number of times in order to prove to Itself that It can do them once; therefore, I don't believe that repeating the possibilities of energy is likely, as nothing is gained by it and God has no 'other' to whom It has to prove Itself.

There is another alternative, as well, one that is, almost unimaginable by humans and that is to create universes with some other substance than that which we call energy and to explore the possibilities of what can be done with that other substance. To my way of thinking, this is far more likely to be what follows the 'end' of the explorations of the 'possibilities of energy'. I believe that it is far more God-like/Creator-like, because it is far more creative, to create a different underlying substance and explore the possibilities of that substance than it would be to continue to repeat, forever, the events that can be performed by energy. As to what that other underlying substance would or could be, I'm sorry to say, is outside my ability to apprehend. I've only ever experienced a universe based on energy and, as no other substance exists in this universe, it is outside my experience as to what other substances would or could be created by an omniscient and omniprehensive entity. To quickly recap, without complete understanding, which is what I mean by 'omniprehensive', complete knowledge, in other words, omniscience is useless - even to a Deity. God completely understands the complete knowledge God has and, for that reason alone, God can act wisely.

So, the point, here, is that the possibilities of energy are not endless; rather, they are vastly huge but finite, as even such a creature as limited in knowledge as I am can think of things that energy cannot do. And, because of that finiteness, the beginning and endpoints of a space-time continuum CAN be, and must be, defined and, thusly, a space-time continuum comprised of energy is teleological. Even if it is immediately followed by a space-time continuum comprised of some other substance; that substance would run its course and, perhaps, be followed by another continuum based on another substance and so forth. Additionally, I suspect that the number of basic, fundamental substances from which a universe can be constructed is, also, limited. I realise that last statement is pure speculation; but, I believe it is no more far-fetched than stating that there must be a limited number of combinations of sub-atomic particles into which energy can conform, and that is an established fact, even if not all of them have been observed.

The point is that there are only so many possibilities that can - and will - happen. Although that number is vastly large, it is finite and that limit is what binds the ends of the space-time continuum and makes the universe teleological even if substances other than energy are exploited by the Creator at some point in time. For all we know, and there would be no way to discover whether or not this is true because no one could ever observe beyond the limits of their own Big Bang-to-Annihilation cycle, the other basic substances have already been fully exploited and we have finally come around to what we call energy. But, truly, I have no idea where we are in the midst of cycles - only that we are, most likely, in the midst of cycles.

... The Mostly Religious Side

The Qur'an alludes to the fact that this is not the first universe created by The Creator. In 'The Spider', Surah 29:19, it says, "See they not how Allah originates creation, then repeats it: truly that is easy for Allah." And 29:20 follows with; "Say, 'Travel through the Earth and see how Allah did originate creation: so will Allah produce a later creation: for Allah has power over all things.'" In 'The Romans', Surah 30:27 "It is He who begins creation; then repeats it; and for Him it is most easy... " While this is not scientific evidence, it is religious, revelatory evidence and, I believe, that we should not ignore it. If this is not the first universe, then it DOES imply that a series of universes is a natural part of our reality and that lends credence to the geometrical conformation I put forward in my model: that of an expanding sphere inside a toroidal backdrop. And, as I have stated earlier, that geometry is the simplest and most elegant way to ensure a sequential series of universes whilst maintaining the whole of the sequence in a space-time continuum. And you would expect nothing less from an omniscient and omniprehensive Creator.

But this forces us to accept a few hard truths. For example, I am writing this particular section just a few days after the 6.3 magnitude earthquake at Christchurch in New Zealand. During and after that earthquake, several people died - some, no doubt, in horrific ways like being crushed by a collapsing building (At the time of writing that sentence, little did I know that the earthquake, a 9.0 on the Richter Scale, and the resulting tsunami were hitting Japan). Knowing the universe is teleological, forces us to accept that their deaths were planned by The Creator from the inception of the universe. However, we must trust The Creator, as we know It to be All-Wise and, therefore, to act in a fashion that is Most Wise in order to achieve Its goals. Therefore, we know that those deaths were NOT in vain; rather, they were part of a larger plan to get the universe from where it was to where God knows it ought to be. And God knows best! Those souls weren't lost, they were saved; but, from our vantage point within Creation, we don't get to see that part. We must remember that conservation of energy is reasonably suggestive evidence for a belief in an immortal consciousness.

On a larger scale, knowing that the universe is teleological means that each and every death is a vital event that is part of God's plan for the universe. And God's plan will not be thwarted. It cannot be thwarted because it is written into the very fabric of the space-time continuum - this is another implication, albeit of a religio-philosophical nature, of Einstein's Theory of Special

Relativity. What is outlined to happen will happen and we had better accept it; thus the importance of acceptance in Buddhist thought. There is simply no rational argument against teleology in the face of Special Relativity. Is that so hard to accept? It certainly simplifies certain things and the simplification of things is, again, what a 'Unified Theory' would do. God says, "Be!", and a thing is. And all those things have already been done from God's perspective; we're just doing our thing in the midst of it all.

We must trust that Our Creator will take into consideration the situations regarding our deaths and ensure that God's Holy Mercy will prevail - especially in cases that we mere humans may see as unnecessary deaths. God cannot NOT know what each of us feels and thinks, as that is the way omniscience is maintained, so know beyond doubt that all situations are considered - there can be nothing but God that is more well-acquainted with what occurs in the universe. In truth, there is no such thing as an unnecessary death and they are all vital to get the universe from where it was to where it's going. God's Mercy DOES prevail and, if truly innocent lives appear to be 'lost', then those souls are saved, as God never tires in doling out Holy Mercy. If you have lost a loved one in just such an event, then, with firm conviction, know that God has taken them back and dealt with them in the Most Merciful way possible, which, logically, would be to admit them to Paradise.

Understanding a teleological universe in this way helps us to trust God and to trust God completely. Why? Because we know and accept that there is a plan to the universe and that our lives, from beginning to end, are a vital part of that plan - yet this can be derived from science through an understanding of Special Relativity without even accepting the premiss of God; however, I feel I've done enough already to demonstrate the likelihood of God's existence but for those still in doubt, Pascal's Wager, too, is a good bet. We know that, even in the face of seemingly horrible events, God's Holy Mercy will prevail, as it logically must, because God is the Most Merciful entity in existence, if for no other reason than that God is the ONLY true entity in existence. All pain that is felt is felt by God; and a God who is 'Most Merciful' would always diminish that pain if it were possible to do so. So fear not; God cannot and will not forget a single one of us.

In this way, understanding the teleological nature of the universe aids our trust in God and, in truth, it should give us incontrovertible proof that no death - even those that appear to be unnecessary - is in vain and that all of our lives are completely necessary and vital to the integrity of the universe. Furthermore, each moment of each of our lives is just as vital as the next and, for that reason, we should pay great heed to every deed we commit, as each act leads the entire universe forward to God's intended goal. Surely, we would prefer to look back on our lives and think that we played, not only a vital role (as each of us can say that), but, a positive role that helped the universe and ultimately caused more good than bad to occur in it on our account.

Note that I'm not screaming for viral rights here or that we should not continue forward with human medicine simply because we can logically derive that bacteria and viruses are vital to God's plan. If you want evidence for that, you can derive that evidence from the fact that we have reacted to them by discovering antibiotics and antiviral drugs; so, bacteria and viruses have served a useful purpose with regards to humanity's self-defence as a species. Just, morally, ignore the fact that we have to fight them only with biological and chemical weapons and live with it. Sure, it's genocide on a massive scale; but, it is self-defence and that's allowed. They are, in a very real sense, enemies that will kill us if we don't kill them first. It's nothing personal, mind, but they are programmed to live by attacking and/or killing us. And watch out for those prions, like the one that causes CJD. Just a tiny snippet of amino acids that sneaks across a cell membrane and finds the first bit of hot-to-trot Transfer-RNA it can, attaches itself, and they have lovely CJD children together; it's madness, absolute madness to allow that and doctors know it and are trying to figure out how to prevent it. It's absolute Hell inside those cellular walls; anything's fair, let me tell you.

Funny how, with a bit of philosophical insight, Special Relativity confers such great responsibility upon us that makes each and every act we perform vital to the whole and, as we've learned from Newton's Laws of Motion, each action demands an equal and opposite reaction. If it is Paradise we seek, then act with great care. But, also, never, even for a moment, entertain the thought that anyone has died in vain or needlessly, as that thought is a betrayal of one's trust in God; and God, if anything, is the only thing you can trust. God has Its reasons, divine reasons, for performing all these acts and the fact that we are neither omniscient nor omniprehensive means that we may never understand the full reasons behind any act - even those that seem completely harmless or inconsequential; but, that does not, in the least, suggest that there are no reasons behind any event or that any event has no consequences. It certainly does not do so in this cause-and-effect universe. The fact is that, for humans, the universe is too complex to apprehend, much less comprehend. Things may look chaotic to some; but, I assure you, if it looks chaotic to you, it's because you do not know where anything is headed. Well, nor do I; but, I can see and point out why no one can.

Special Relativity confirms that there are no inconsequential events, as each event is vital to the entirety of the space-time continuum. Therefore, it is fair to say that 'every act is a sacred act' or, perhaps more properly: 'every act, in the mind of an enlightened individual, is a sacred act', because the enlightened individual knows that every act is vital to the whole and they know that it would be in their own best interests, and in the best interests of others, to act in the most responsible way possible, as they would know that, by acting in ANY way, they affect the entire universe.

Finally, the philosophical debate regarding teleology is at an end (pun intended!). It SHOULD have been realised to be at an end in 1905 when Einstein published his Theory of Special Relativity; but, due to the stubbornness of some scientists who preferred to avoid the philosophical implications of the theory, it has dragged on for over a century longer. Perhaps they thought we can't handle the truth. I don't know. Nevertheless, please, now understand that the universe is teleological, irrespective of what any scientist may say to the contrary, and that God has a plan for the universe and for each of us literally woven into the fabric of space-time. Understand that each of us is vital to the whole and that our existence within the space-time continuum is both conclusive and scientific proof of that fact. And understand that we should completely trust God's judgement and actions - even in the face of what may seem to be injustice. God is omniscient, omniprehensive and acts with perfect wisdom towards goals only God fully comprehends. Trust in God completely and know that Einstein's Theory of Special Relativity stands as scientific proof that you ought to.

Are Newton's Laws of Motion Purely Physical?

This thought came to me after remembering that Sir Isaac Newton was, first and foremost, an alchemist. It was very much in his nature to be as holistic as possible in his approach to science. But he also knew that the Scientific Community, even in his time, was dubious about spiritual matters and he wouldn't have wanted to risk his authority and respect by speaking plainly about his beliefs. So he cloaked them to the extent he felt required to do. Each of the three laws of motion refers to 'bodies in motion' and each is believed to refer to physical bodies in motion; but, what if Newton had intended to include spiritual bodies as well? There would be no need to change the phrase 'bodies in motion' as spiritual bodies could be included without being particularly specified. But is there a spiritual corollary for each of Newton's laws of motion? Shall we investigate?

The first law states that "unless acted upon by a net force, a body at rest stays at rest, and a moving body continues to move at the same speed in the same straight line (direction)". Given my background in having studied comparative religion, when I re-read that law in light of trying to take it to include spiritual bodies, it just screamed out to my mind, 'That is the Western, scientific version of the concept of the gunas of Hinduism'. The three gunas are: Sattva, Rajas and Tamas. They are spiritual qualities or forces that, together, express the 'net spiritual forces' that affect us. Sattva is usually depicted as simple, clarity of mind; Rajas is a disruptive, disturbing influence and Tamas is dullness and lethargy. In this analogy, I see Sattva as representing an individual's truest sense of self - their own unsullied consciousness; Rajas (the general disruptive, interactive force) and Tamas (spiritual inertia) combined would be how one individual experiences another individual's Sattva. Whilst it is true that one can be affected by another's Sattva, it is harmonic enough as to not distress the soul like the other forces of Rajas and Tamas do. Tamas is what keeps a depressed person depressed and why it's harder to motivate a depressed individual than one who is not depressed. So, too, a mind/soul filled with Tamas will tend to remain at rest (and depressed and slothful and, in extreme cases with the right combination of Rajas, self-harming) until acted upon by sufficient Rajas (and/or Sattva [but it takes more Rajas at first!]) such that it can, once again, achieve its own Sattva. Too much Rajas can make an individual aggressive, like a bull in a china shop and is what keeps the manic, manic. Sattva is the quiet forward motion with no external forces impinging on it. Too much Sattva usually leads to moksha, the deliverance and release of the soul from the cycle of re-incarnation that is the goal of Hinduism, and is not considered problematic.

So, to paraphrase Newton's first Law: A (more) Tamasic soul will tend to remain Tamasic until acted upon by Rajas (and/or Sattva) and a (more) Sattvic soul will continue to be Sattvic until acted upon by Rajas (and/or Tamas). I inserted the word 'more' there to denote that each soul is, in most but the rarest of cases, comprised, to some extent, of all three gunas. Here we have a sound spiritual concept, which had been recognised by Hindus millennia ago, that is an almost perfect corollary to Newton's first Law.

Looked at another way - probably Newton's alchemical way - Sattva becomes Salt, that perfect combination of opposing ionic elements that forms a complete bond with itself (its Self). Rajas is Sulphur, the fast burning element that scorches its way disrupting and disturbing. Tamas is, then, Mercury, the heavy, liquid and poisonous metal. I think Newton understood the gunas in this way and may well have hinted at it in this first law.

Newton's second Law states that "a net force applied to a body gives it an acceleration proportional to the force and in the direction of the force." This is vastly important. Given the first paraphrased law, this second law implies that the interactions between spiritual bodies impart an eternal effect, that is, when one set of gunas (one spiritual body) communicates with another, it imparts a force that is irremovable and it receives a force that is irresistible. On a spiritual level, every interaction is the meeting of irresistible forces. From that moment forward (in a spatio-temporal cone), all the actions of B have become affected by B's communication with A and vice versa. Spiritually, we can interact in an intellectual and/or emotional way with one another, not to mention that intimate, physical communication, certainly, can have emotional effects. I believe this is the 'emotional communication' that Gregg Bradon intended in his book 'The Divine Matrix'; especially his 'Key 4': "Once something is joined, it is always connected, whether it remains physically linked or not". This concept is also related to the concept of "one flesh" in a Christian marriage ceremony, where the two individuals are said to become "one flesh". I firmly believe that reference is metaphorical and represents that, once two people have had 'spiritual intercourse' - a topic for discussion all on its own, but I mean it in its simplest level of even including a casual 'Hello' passing by someone on the street, which is, indeed, a simple form of charity - they have a permanent spiritual bond, as the effects of their interaction continue onwards throughout space-time. Also, the implication that our interactions cause eternal effects from that point forward should make us feel the utmost responsibility in just how we impart and/or receive one another's influence. So, to paraphrase Newton's second Law: An individual soul always imparts gunas when communicating with another individual soul. Here we have another sound spiritual concept of which the Hindus have been aware for millennia. Alchemically, it's simply that all spiritual interactions can be reduced to the principles of Salt, Sulphur and Mercury - the absolute foundation of alchemy.

The third law is the one most of us have already derived or run across at some point but, for completeness' sake, I have to discuss it. The third law states that "When a body, A, exerts a force on a body B, B exerts an equal and opposite force on A." This is sometimes phrased, "for every action there is an equal and opposite reaction." To me and countless others I've met, this stands as a corollary for karma, as our karma is our collection of gunas achieved over our life through our interactions with the universe, especially, though, living things. On a spiritual level, one's karmic debt, in Hinduism, may force one's soul into a particular transmigration or release it completely in moksha; whereas, in Western religions, one's soul, due to its spiritual attainment (roughly equivalent to karma), is rewarded or punished in Heaven or Hell. Both of these explanations serve as spiritual corollaries of Newton's third Law. Alchemically, it could be said that beneficial spiritual interactions can turn base metals into precious metals and malicious spiritual actions can turn precious metals to base metals. I'm sure there must have been an inside joke amongst English speaking alchemists about turning base mettle into precious mettle considering that the usage of both terms was common in Newton's day. So, perhaps Sir Isaac was trying to tell us about all bodies in motion—not just physical bodies, but spiritual bodies, as well.

Do We Have Fate or Free Will?

Although I've already covered this in the chapter, *"Is the Universe Teleological?"*, I stated there that I would cover it more fully later and so I will. Here, again, is another one of those parts of the book where some people are going to want to throw the book against the nearest (or farthest!) wall. If you do and see me later, let me know and I can say, "I made you do that!" My angle on the age-old question of whether or not free will exists is strictly down to the direct philosophical implications of understanding Einstein's Theory of Special Relativity, which I've alluded to many times before; however, I need to give it its proper fullest attention. The short answer, which will, no doubt, upset many, is that we don't have free will. The slightly longer answer is that we don't have free will as we normally think of it. Now, that last answer is very mysterious and begs explanation and it is in that explanation where, I hope, you will understand why we have to accept a new way of viewing free will. But, firstly, you need to understand why Einstein's Theory of Special Relativity forbids free will and that explanation leads us nicely into one of the basic precepts of this book that 'free will is not what we think it is'.

In 1905, Albert Einstein released his, now famous, paper on what he termed 'Special Relativity'. He had theorised that space and time were not separate dimensions but, rather, were different aspects of a larger 4-dimensional whole that was the 'Space-Time Continuum'. This space-time continuum was an object that contained all of space and time - all of it - and that, in a similar way to how the 3 spatial dimensions are joined together to form a 3-dimensional space, the truth was, in fact, that time itself was also joined to the three spatial dimensions to create a 4-dimensional reality. His theory made certain predictions about what would happen to clocks that moved through space-time at different velocities and, hypothesised that, for clocks that were moving relatively faster than others through space-time, time would appear to slow down for those faster-moving clocks. This has, since, been demonstrated countless times and has always shown to be true. It is known as time dilation. In fact, our modern GPS, which relies on clocks that are in geosynchronous orbit around Earth - thus moving faster than the Earth in order to stay directly over a particular point on the surface - account for the ever-so-slight-deviation between their time and our Earth-bound clocks and that is why GPS systems are as accurate as they are; if they did not account for that difference, it would only be a matter of months before someone in St. Louis, Missouri found their GPS telling them they were in Majorca. Therefore, because the theory has been proven by these time-dilation experiments, the theory has been accepted as fact and science now readily acknowledges that the universe in which we live is, indeed, a space-time continuum in which time is equally joined to all three spatial dimensions and that it forms a greater, 4-dimensional reality.

At first glance, this may not seem that big of a shock, yet, in 1905 it caused quite a stir. For one thing, it implies that there is a single object, called 'the space-time continuum', which contains ALL of space and time. This means that there is a single, 4-dimensional object that has within it all the past, any given present and all the future. In other words, the future is already extant in the whole of space-time. The future already exists! This easily leads us to the conclusion that free will cannot exist 'as we understand it', or, perhaps more properly, 'as we understood it', because all our future events already exist and are just waiting for space-time to expand to that space in time where we will do what is destined for us to do at that time at that place. The discovery of Special Relativity uncovered the answer to the age old question regarding free will versus fate and fate wins; although it was not Einstein's intention to end that philosophical debate, nevertheless, he did and should have been given some kind of prize in philosophy for it, as well.

I'm sure this comes as a hard thing for many people to accept that, in reality, the choices they make have, in effect, been laid down and fixed in the fabric of space-time since the Big Bang. They will insist that it's obvious that free will exists because they can choose to do either A or B. Well, it's also 'obvious' that the Sun moves across the sky; but, that doesn't mean that the Sun revolves around the Earth, as Copernicus proved. Although the Roman Catholic Church put up an almighty fight against Copernicus' 'truth', eventually, they conceded. Especially once it was realised that the concession didn't really remove any power from God. I'm afraid that now it's time we finally concede the truth regarding free will. Yet, how do we account for the appearance OF free will?

I believe this boils down to two distinct facts:

1) **We have no access to the future like we do to the past (with memory).**
2) **We have the ability to speculate about the future at any given (present) time.**

These two facts conspire together and create a "perfect illusion" of free will. Given an apparent choice between A and B we can speculate that we can do either; however, when it comes time to actually act, we can only perform one of the two acts and THAT was the act that had always existed in the space-time continuum. Alternatively, if we are provided with a multitude of choices, say ten, we can still only perform one of them at any one time at any one place and the 'choice' we make is the spatio-temporal event that was, is and always will be in the space-time continuum. The fact that we can speculate about what MAY be is completely irrelevant to the act we make insofar as we can only ever perform one act at any given time at any given space. These speculations are, in fact, simply other spatio-temporal events themselves, although they are mental events, they are, nevertheless, spatio-temporal events because of the changes to the 4-D-to-Consciousness interface. The fact that we cannot see into the future like we can see into the past with memory casts a veil across us with respect to future events and we are blind to them. If we could see into the future, we would know exactly what we would do and we would know that the future is out there just waiting for us to 'catch-up' with it.

The fact that we are veiled to the future is, though, a very good thing, although some may say that we could prevent disasters from happening if we could see them, you would find that, if you actually saw them happening in the future, then they WILL happen and you cannot avoid them. You would find that you were prevented, somehow, from altering what WILL happen, otherwise, you wouldn't have seen it happen, if you could, truly, see into the future. The reason that I say it's good that we are veiled from the future is two-fold.

One, it prevents us from being bored with life. If we always knew what was going to happen, we would just go through the motions and not really 'live' life. There'd always be situations like: "I knew you were going to say/do that!" and "I knew that would happen." In fact, life would be nothing short of a long chain of these I-knew-that-would-happen events. Life would be absolute drudgery and unbearable, certainly with regard to tragic events that we KNEW we couldn't prevent and would be forced to, perhaps, watch helplessly. Two, it places responsibility back into our own hands. There exists an argument that, if the future is all laid out for us then we are not really responsible for what we do. Well, because we don't have access to the future, yes, we ARE, then, responsible for our own actions because of our ability to speculate before we act. We have that moment of time to speculate about what 'choice(s)' we make and the amount of consideration we make during those times of speculation bears relevance on our personal responsibility for our actions. Given a scenario where we could see into the future, we know that we will pull the trigger and shoot the unarmed man, thus we know that we're a murderer long before we've done it; however, if we are veiled from the future and do not know the outcome, right before we pull the trigger, we may think, "Wait a minute. I don't HAVE to do this" and 'choose' not to pull the trigger, then we are NOT a murderer. The fact that we don't know the future forces us to consider our actions and that puts the responsibility firmly back onto us.

This is what I mean by a 'perfect illusion'. It is completely inescapable and, for all intents and purposes, we are still left in a situation where we bear the responsibility for our own actions even if those very actions have been, allegorically, etched in stone beforehand. In my mind, only a very clever Creator would create a universe in which the reality is that the entire universe is completely set up from beginning to end, yet the most intelligent creatures on this planet, humans, are still left in a position where they can be held accountable for their actions. This, indeed, is an extremely clever and wise construction and something that you would expect from a Creator that is omniscient, omniprehensive and omnisentient, that is, all-knowing, all-understanding and all-wise.

I feel that, for the most part, this book will be received by a large number of people as stating one thing and one thing only - that humans do not have free will. My answer is that the argument for individual free will was refuted forever by Einstein's (well, Einstein's discovery that our space-time was a 4-D Minkowski space. Minkowski also understood the implication with respect to free will) concept of space-time and, when he was asked about it, he responded that "Free Will and Space-Time are not compatible". Because space and time are intrinsically linked together, the whole of space-time can be viewed as a single object containing the entire past, present and future of every existing quantum of energy. In this space-time there are no missing points. The future is just as much set in the fabric of space-time as is the past. The difference is that, because of how our consciousness is set up, we can recall the past but not the future. If you thought you had free will, you've been duped. God is omnipotent and is not allowed to remain omnipotent and allow any will other than its own.

Einstein didn't dwell on the concept (the loss of free will), though, and I believe this for two main reasons: 1) He was a physicist and not a philosopher, and 2) He didn't want that 'fact' to detract from the great discovery in physics that he'd made.

But I say that it's finally time to understand the philosophical implications of Einstein's theory and, through an exploration of them, come to a new understanding of our role in the universe. I don't want people to walk away from reading this book with the feeling that I have robbed them of their free will. If you want to blame someone, blame Einstein back in 1905. I simply wish to stress the great responsibility that is placed before us because of the appearance of free will.

As I said, I don't want you to feel robbed of your free will. No one has taken it because, in reality, you never actually had it; but, what you had - and still do have - is the perfect illusion of it, which is, for all intents and purposes, just as valuable and as effective as the real thing. God can't allow you to have actual free will because Its will must prevail. God is omnipotent and allowing humans (or Jinn or caterpillars or gazelles or, indeed, any creature with an awareness complex enough to interact intelligently with its environment) the actual ability to act in defiance of God's will would mean that those creatures would have power that God does not have. If God is not in total control, God is not omnipotent. Therefore, God cannot give you will. God can, though, give you a slice of Its will and, in fact, this is exactly what I purport we all have.

We are all 'agents of the Lord' and act in complete accordance with God's will. We may, at times (sadly all too much of the time), act against God's guidance; but, we cannot act against God's will. If we could, logic dictates that God is not omnipotent

176

and God must be omnipotent. Our acting against God's guidance is, by definition, sin. The retribution for sinning is, of course, the equal and opposite reaction to our sinful acts. But only God knows what those are for sure - unless God has 'revealed' them to a prophet, which is purported to be the case with respect to certain sins. We may impose certain punishments upon ourselves or others in this life but God will impose perfect justice upon us ultimately. Each case will be different and, in each case, perfect justice and perfect mercy will prevail. If we have earned time in Hell, it is completely upon our heads. If we have acted carefully or if God's mercy permits, we may enter Paradise. In any of the possible outcomes, the result will be because of the interplay between the physics of 'equal and opposite reaction' and the dispensation of God's justice and mercy; but the mercy precedes the wrath.

Please remember the above and NOT that I have taken free will from you. It's not for me to take that which you didn't have in the first place and it is not my desire to depress you emotionally or cause you to feel a loss. In a universe where, truly, there exists only one thing, that thing is responsible. The other trick that coincides with the perfect illusion of free will is that each of us is an extension of that one thing that exists, which makes us responsible. We are, in essence, blind agents who can't see what they're about to do; but, because of that blindness, the responsibility for our actions rebounds back to us. So, we need to be careful. I will take responsibility for pointing out the philosophical ramifications of accepting Einstein's Theory of Special Relativity, but I didn't design the universe; so, please, don't shoot the messenger! I tell you the truth because I believe you deserve it and can handle it; and, so that you can take greater care in your actions in order that you will merit Paradise. To have moved people closer to Paradise by leading them to an understanding of the truth regarding their plight and place in this universe, that is my sincerest hope.

After reading this, you are just as responsible for your actions as you were before you read it; the difference will be that, after reading this, you will know exactly how and why you are responsible for your actions. It is that education that is my goal and, of course, to ensure that people take advantage of that precious time we have before we act, so that when we act, it is with due consideration with respect to our future and the future of others both in this world and the next. We must think of others' rights and act in such a way, if at all possible, as to protect the souls of others as we would have others protect our own. This is the reason why we should 'do unto others what we would have others do unto us'. This understanding of free will is what proves the concept of Al-Qadr (fate, destiny or kismet) in Islam and the concept of Divine Providence in Christianity and Judaism without undermining our own individual responsibilities. With Special Relativity, Einstein not only came close to discovering what he intended, which was to discover 'how God created the universe'; but, he also came incredibly close to uncovering one of the reasons why God created the universe in the way It did.

What about Emotion?

Emotion, i.e., the emoting or expressing and demonstrating of our inner feelings to others, imparts an effect on another's consciousness. Deriving from the spiritual take on Newton's Laws of Motion, we can plainly see that it is through emotion, the process of emoting, that is what demands an exchange of spiritual forces (an exchange of gunas); one person acts and the other reacts and the first reacts to the reaction of the second and so on. And because, in my model, consciousness exists in the Calabi-Yau space (which, in effect, is like a *Star Trek* 'sub-space' in that, because it is outside of space-time, communication is instantaneous due to the fact that, once an exchange has been created [that is, once space-time has expanded to the point where that exchange of emotions is taking or has taken place {or even a thought being thought and thus, equally, a feeling - an inward perception - being felt}], it can be retrieved at any time - although the process is slowed down by the physical interface), emotion, as effective a motivator of energy as physical motions, can impart instantaneous effects via natural quantum entanglement on all that which is and every emotion we make will have a ripple-effect on, at least, the entire future of space-time.

Because we have been placed into a universe - without choice - but been given no obvious means of knowing what will happen (I say 'obvious', as I'm ignoring the concept of divination for this argument), we are faced with the appearance of many possibilities, as we have also been given - without choice - the ability to speculate. Well then, let us take the time we have to speculate, as our greatest hope of knowing has always lain in our ability to guess correctly (scientists call that hypothesising, once they've proven their guesswork); but, we should always try not to

179

rely on an uneducated guess if there's a chance of learning first and as a result, relying on more educated guesses. This is what yields results when standing on the shoulders of giants.

As I've mentioned above, it is these two factors: the inability to know the future coupled with the ability to speculate in the present that lend us - nay, forces upon us - the gift of the perfect and inescapable illusion of free will. But feelings are inward perceptions that are felt by us uncontrollably. When someone derides us, we feel hurt and we can't help feeling hurt; it boils up from within us. In *Star Trek*, Vulcans strove to control their emotions but they knew it was completely illogical to even attempt to control their feelings. We have feelings as inner reactions to our environment; they are, simply put in terms of physics, part of the effects of nature - our environment, and that, of course, involves all the interactions of humanity - on us. That which we do have the ability to do, though, is to take advantage of the time we have to speculate on how we express our feelings and THAT is the power and meaning of emotion. An emotional reaction is a form of intelligent reaction and our reactions ought to be as intelligent as they can be.

And, we must react emotionally in the knowledge that we all are, in fact, an indivisible extension of the 'One Thing' that is the only REAL actor in the physical system that I put forward. Note that, heretofore in this chapter, I have not used the word 'God'. It isn't required that an indivisible extension of the 'One Thing' acknowledges that Thing as God or even that an indivisible extension acknowledges the Thing at all. Certainly, there are things that exist that demonstrate no awareness of their environment; we commonly refer to these things as non-living, as living things demonstrate an awareness of their environment and react 'intelligently' to it. Thus, the rock is left barren to erosion but we can clothe ourselves and hide from the wind and rain and the elm can shed its leaves to protect itself from the coming cold that it knows is coming because it has perceived, through chemistry -

pure chemical intelligence - alone, the lessening of light. The acknowledgement of God's existence in all this isn't required for the system to work, thus the position to speculate that this One Thing should or should not be called God does not prevent the system from working; rather, it becomes just another cause for effects that continue to have effects.

Does The One expect the mountains to bow down to Him? Not exactly - but they DO insofar as, over time, they are reduced to sand on faraway beaches. And if all we did was to spend our time bowing down to God - and God has never required 24/7 awareness even of Its most devout believers - we would be left to the ravages of erosion and be as effective as the rocks. Rather, what is required of us is to do that which we will do. And some of us living things will take time to acknowledge the Thing as God and others will not. That is a perfectly natural and normal thing to expect. Thus, an atheistic view, that is, 'a non-acknowledgement of God', is a thing that one would expect to see in the universe - even a universe in which God exists. And there is simply no sound reason to think that that an acknowledgement of God is required from all living things, so I wouldn't expect even all humans (or any creatures capable of conceiving OF God) to acknowledge the existence of the God. Rather, what I would expect and sincerely hope for is that humans would acknowledge their OWN existence and consider it, for a moment, in the knowledge that we are an indivisible extension of the One Thing that exists - call it what you like.

All of space-time led up to our birth. And we are thrust into the world to do that which we will do. And we humans are given the gift of speculation, the ability to see possible futures. And, of course, we all have our own desires, hopes and dreams. If we are to be the very best that we can be and do the things that we really would prefer to do, then we MUST act in such a way as to best afford that. And, whilst doing THAT, we should also acknowledge one another as indivisible aspects of the One Thing. We should take the time we have to speculate on how to best interact with the entirety of our environment, which, naturally, includes all of humanity. It should be felt as incumbent upon us to act in such a way as to foster all of our greatest hopes, desires and dreams. Equally and dreadfully, this is the most selfish reason that lead me to write this book; I felt it as a duty to my fellow humans.

The homeless, drug addict, half-dead on the park bench, who has been forgotten by humanity, still exists! And he is waiting there to be saved from his plight. His hopes and dreams have been laid aside by us for we do not take the time to consider him in our busy lives - and our lives ARE busy. Would it not make sense then, for someone to take the time to ask that man what it is that he wants out of life? Is there no one who cares? Perhaps it is the man himself, languishing and anguishing on the park bench right under our noses, who would rather be helping people in just that way. I sincerely hope that, by considering that we are all vital to the whole, as evidenced by our existence, that we come to the understanding that we cannot forget one another. We MUST consider one another. Furthermore, as an ethic derived from the physics alone, we should act in such a way that fosters to our best abilities, all aspects of the One Thing, otherwise, our inaction acts as a form of disease that eats away at aspects of the One Thing, as evidenced by the poor, starving, homeless, beaten, oppressed, etc that could be eradicated if we cared enough to organise ourselves to solve it - not to mention our impact on our planetary environment.

Our emotions are the key to solving this very problem. Can we emote and demonstrate care regarding one another? Exactly how we react to our environment causes permanent changes to, at least, the future. This is pure cause-and-effect physics and no acknowledgement of God is required, rather, only an acknowledgement of that which exists is required. And there are poor that exist, and those that starve, and those that are homeless and those forgotten in other ways. And it is through our emotions, through the very way we react to our feelings about, i.e., our perceptions of, our environment, that we can change the future forever.

Simply put, emotion is vital. It represents a two-fold effect in my system. On the one hand, there are the obvious effects that are projected when we interact and those are vital to shaping the future, about which we know nothing other than possibilities. On the other hand, there is that sub-space aspect of near instantaneous effects when one's consciousness effects another's consciousness - and consciousness is in the Calabi-Yau space outside of time - and those effects are everlasting, as they exist outside of time. And if consciousness does really exist outside of time (by virtue of being outside of space-time), as I purport, then those effects could even, potentially, affect the past. Thus, by remembering only the good things, that could, potentially, actually alter the bad things that are forgotten and move them from 'the set of things remembered' to 'the set of things not remembered'. This weird emotional yet mathematical operation of set theory is the physical mechanism behind forgiveness. Mathematically, the operation involves multi-dimensional calculus across the Calabi-Yau space and is out of scope philosophically; nevertheless, I felt it should be mentioned to those who can see the relational algebraic aspect that forgiveness represents. When we forgive and forget, we remove the 'bad feelings' from our consciousness. And, as I've said, our consciousness is just one slice of a larger integrated consciousness that is the consciousness of the One.

Now I've brought the mathematical concept of Set Theory into play and I think it's a reasonable analogy. It is simply through Set Theory that people can be forgiven. Forgiving another for certain actions, which may have been reactions themselves, is done through our emotions. In order to actively forgive, we must let the 'other' know that we have forgiven them. We must express to them our inner feelings of forgiveness; we must emote it to them. It was for that reason that the line in the Lord's Prayer reads the way it does: Forgive us our debts AS (i.e. 'whilst and at the very same time') we forgive our debtors. There is only One Thing that exists, and we are all extensions of that Thing. When we forgive, we are forgiven, because there is only One. When we forgive another, that One (for it is only that One that exists) is forgiven. But we are not forgiven until it has been emoted to us nor have we forgiven until we have emoted it.

Jesus was employing theological teachings in common speech when giving his discourse of the Lord's Prayer, but, in effect, he was also making a very simple statement about the physics of the system in which we exist. I believe he understood the nature of the oneness of the universe, albeit derived from an abiding faith and/or a personal revelation from God rather than an empirically demonstrated aspect of the known universe such as the space-time continuum. But, now, the evidence of the space-time continuum - empirically proven by the demonstration of time dilation - alone is evidence that forgiveness could work in the aforescribed manner, even without extending consciousness into the Calabi-Yau. Even given just the Standard Model and accepting that consciousness abides purely within the 4-D space-time continuum, there is only One Thing that exists - and it IS that continuum, irrespective of how many other dimensions an individual admits. And, if forgiveness is, as I suspect it is, stated in physical/mathematical terms: a function of conscious set theory against 'the set of things known' that moves 'the set of our bad feelings regarding X' into 'NOT (the set of our bad feelings regarding X)' via their negation, then we can change the past through forgiveness.

It is, then, through our emotions and forgiveness above all others that, perhaps, we can change the past. And no one has ever known how to change the past before. But, there's a part of the Standard Model that says that forces SHOULD extend in all directions: the old 'equal and opposite reaction'; but what is the opposite of the present, in which forgiveness takes place? The answer is: both the past and the future. Then, if space and time are joined as they are, the past should lie just as exposed to 'present' forces as the future is. But no one, before now, could ever pinpoint a mechanism that could do it. Not until emotions crept into the laboratory. Note, that on a more technical level, space-time still has to expand to the point where an act of forgiveness takes place and only THEN does the ripple effect have its effects on the past. Whilst, in a larger sense, it may be true that the past once changed has always been like that; if so, then a seeming conundrum arises. But in the Standard Model of the universe, there has always been the argument that forces act in ALL directions - and space and time are joined. So, equal and opposite reactions are required to any force applied and any full and reasonable science MUST include acts of consciousness. So now we see a methodology, a process that can take place on the level of consciousness that can meet the expected 'backwards flux' that the Standard Model proposes is possible and should occur, yet for which modern science can show know evidence. Well, of course not, consciousness only acknowledges that which it acknowledges and forgiveness is a disregarding OF acknowledgement, so the evidence disappears. But the Standard Model doesn't address consciousness except for the specious 'to the extent that we are aware, we are conscious of that which is called the Standard Model'. Well, now it has a far less specious and cursory treatment.

When we forgive, we no longer hold those bad feelings and we disregard them; and, in so doing, 'the consciousness that is' disregards them. Thus forgiveness is, perhaps, the most powerful of all emotions, as it seems to be, of all the emotions, the one that is most likely to affect both the future and the past by virtue of how it changes our conscious memory of the past. If we have no claim via memory to 'bad feelings regarding X', then they are gone. The entire process is, in reality, mathematical and doesn't require the individual to recognise the One Thing as God. Recognising it as "the space-time continuum", of which we are an incontrovertible part (irrespective of my view that we are, in fact, also an indivisible extension) is all that is required - and that has been proven empirically. So an atheistic view can still see the value and power of forgiveness, even strictly from a mathematical analysis of consciousness and applying set theory to it. And, hopefully, one can see that it has the potential to not only alter the future because we have put our 'bad feelings regarding X' away, but also, potentially, to change the past by changing how we remember what happened and, perhaps, viewing it - however horrible it may be - in a larger context such that we can see its role in our development.

But that can take great emotional maturity and not each of us has been given that. Perhaps emotional maturity can come to us all through a realisation of our role in the One Thing that exists, whether one views that as a monistic and panentheistic God, as I do, or simply as the space-time continuum of which we are an incontrovertible part. And I sincerely hope that these words can lead us all towards that realisation and that we can grow together, as the One we are. In order for us all to move forward in the best way possible, we must all stop to consider one another in the process and act in an empathetic and compassionate manner towards those around us, as if they were no different than ourselves. And by that last statement I mean that we take the view that we are like individual cells of a larger body that is humanity (at least, if not taking the higher view of being an indivisible extension of the One Thing) and treat each other, based on a firm knowledge of physics alone, with the same level of 'fairness' that our own 'healthy' cells within our body treat one another. This is simple courtesy and respect for others; the basis of which forms the ethics for most decent atheists. However, there are plenty of humans that tend to act more like a cancer or AIDS to one another in petty attempts to fare better through competition and lose sight of the larger body. It is my hope that, if my theories are accepted and taught, that the generation that is brought up with this knowledge will no longer choose to 'fight or take flight' but would prefer to take the time to stand and understand and cooperate, seeing that, not only is it easier, but everyone stands to gain (pun intended!).

With any luck (he says, knowing that 'luck' is simply a perception that events have occurred in accordance with our own desires), the generations brought up with my theories, I believe, stand a chance of allowing mankind to survive long enough to see a larger part of the universe than just this planet and can grow to become an active and desirable 'group' in our galactic neighbourhood, if, indeed, we are not alone. If we are alone, then we can grow in the safety of knowing we are, no longer, a threat to ourselves and our greatest enemy but, rather, a friendly and considerate sort.

How Does This Change Justice?

Truthfully, it changes justice radically. When we accept that no individual has free will, as we are required to do when we accept Special Relativity, and that people are destined to do that which they will do, we must accept what they have done after the fact. We must accept that whatever 'crimes' are committed were, in fact, a part of God's will and that the individuals who commit 'crimes' are the perpetrators of God's will; but, simply because we accept that fact does not mean that we approve of the behaviour. Are we then disapproving of God's behaviour? In a way, I suppose we are and in another way no, we are disapproving of the fact that, prior to the criminal act, the perpetrator of a crime failed to speculate that they could have acted in a non-criminal manner. It may seem the greatest of technicalities and perhaps it is, but we must accept it nonetheless.

So, our justice systems are required to change in such a way as to punish based on a perpetrator's failure to speculate that a non-criminal behaviour was possible and that the non-criminal behaviour would have been more beneficial to the whole of the populace. This is, no doubt, a radical alteration of our understanding of justice; but, at least it is congruent with scientific truth. Would we want our system of justice based on a complete misunderstanding of the human condition and, essentially, founded upon lies? No, I suspect we would not, given the choice (and, in truth, no choices ARE given); rather, no matter how bizarrely the truth perplexes the execution of human justice, it is better to work from a standpoint of knowing the truth than forging ahead based on a lie or misunderstanding.

Previously, our justice systems were based on the fact that humans had free will and we punished humans based on their

189

choices; but, we know, from Einstein's Special Relativity, that free will is an illusion and we cannot afford our justice system to be based on an illusion. We must come to terms with the fact that many events will occur that will seem criminal to us but they are, in fact, required in order to get the universe from where it was to where it is headed and God directs that flow of events and, with it, all action. We know that God will enact perfect justice in time; but how can we ever approach anything like perfect justice in our time? I seriously doubt we ever can; but that should not prevent us from trying. The fact that humanity attempts justice is one of our greatest achievements and an achievement that truly can be classed as a thing most Godly.

Remember that acceptance of crime does not mean endorsement. We are led by our consciences to require some form of punishment for an act that the people have decided is unlawful. This, of course, is only applicable to secular nations where governments have enacted laws that govern the conduct of their citizenry. Islamic countries employ Shari'a law, which, by definition, is supported by God's revelation of the Qur'an, and this is what gives their laws and punishments internal authority - the accession to the authority of God as outlined in the Qur'an - rather than an accession of the people to a governing body. Of course, Shari'a law is not always implemented properly and those nations that do not implement it properly will discover, in due time, that their Shari'a law is not supported by any authority - certainly not God's. Nevertheless, secular nations have no outside authority and must derive their authority from the populace, if democratic or the judgement of a single ruler if the country is a monarchy or a dictatorship. It is, in my opinion, dangerous to impart such authority to a single, fallible human and, therefore, I believe that most (not all, I leave plenty of room for Plato's 'Philosopher King') monarchies and dictatorships are doomed to fail justice even if the ruler has the best of intentions.

Some may think, because of what I've stated above, that Special Relativity can be used as a perfect defence enabling the defendant to claim that they were only doing that which they were bound to do; however, this is not the case because of the perfect illusion of free will that is afforded by Special Relativity. The only real change is the basis for punishments in that they must be based on an alleged offender's failure to speculate in a non-criminal manner and, of course, the more difficult aspect of accepting that which has been perpetrated. Buddhist readers will understand perfectly what I mean by the importance of accepting that which has occurred. We must face reality and it is as simple as that. If we turn our backs to reality, we are surely lost. Acceptance is the key but that does not mean that acceptance is easy. Rather, it is, I believe, the most difficult aspect of understanding the universe as it is.

How can we simply accept that countless millions HAD to die in brutal ways? By understanding that we are not omniscient; we do not know the ultimate outcome of any act. The ultimate result of any given act may take place 30 minutes later or 30 years or 3000 years. There is no way to be sure that we, in our short lives, have seen the ultimate result of any event. That understanding is of the utmost importance. Our time here in this world, is very short indeed and, when we see an event, it is almost assuredly in a limited context. We need to understand that our lives are, in most cases, too short to evaluate the actual result of any act. And this is the key to acceptance of that which has occurred. Also, of course, we cannot alter that which has already occurred, except perchance by forgiveness, as alluded to above. To refuse to accept an act that has occurred is akin to turning a blind eye to reality.

Human justice is and has always been only an attempt at justice. Any legal professional will tell you the same. Only an omniscient, omnipotent and omnipresent entity could ever deliver true justice. So, God is the holder of the scales of justice. Our feeble attempts, though, are not unimportant because they demonstrate our intentions towards justice and our apprehension of the concept. God knows that and rewards those who truly strive to ensure a close approximation of justice. Now that we know the reality of the situation in which we are placed, we are further armed with a greater understanding of our role with regard to universal justice. Whether or not the truth of our situation makes our attempts at justice easier or more difficult may well depend on the individuals employed in ensuring it.

This leads us to the importance of ensuring that our lawyers, solicitors, barristers and judges are, in their own manners and activities, as unimpeachable and as honourable as they can possible be. If the interpreters of our laws are acting unlawfully themselves, we are in great trouble. They are our protectors and, as such, we are forced to trust them; if we cannot trust them, then we are forced to protect ourselves from them and this could lead to great civil wars, thus the importance of honour among the legal professionals. The very same holds true of the executors of the law, the police. If they themselves are untrustworthy, we are, again, led down the pathway to civil war as the people must protect themselves from the poor executors of law. Thirdly, the legislators, the makers of laws, must also be impeccable or, again, the people are forced to protect themselves from those who would create unlawful laws.

If the makers of law, the executors of law or the interpreters of law are, themselves, unfit for purpose, the people have the right to protect themselves from the inherent threat posed by these false machines of law and their machinations. For The United States of America, the Declaration of Independence laid down this right in no uncertain terms. But, of course, it is not extended to other nations; however, the way it is written, it implies that it ought to be. I say that, by accepting the reality of our lot within the universe, this right should be extended naturally to all peoples in all lands. When the general populace of the world understands the principles in this book, it will, by conscious osmosis, understand their right to take exception to corrupt governments and their perfect right to act in a way as to protect themselves from the injustices imposed upon them. When governments act in such a way as to reject or disfigure justice or the attempt of justice, they give up their right to govern. It is a fact that when a government acts in a way as to prevent or impede justice, those acts form the root of oppression and no one is required to submit to oppression.

I propose that, if it is noticed by the international community that any government is acting to oppress their people, the international community has a duty to rescue those people from their oppressive governments. And let it be stated that some of the most powerful governments in the world are, today, acting in just this way. Does the international community have the guts or 'chutzpah' to challenge these powerful nations? They have a duty to do so, but do they have the courage to do so? There are, at least, two governments, which have permanent seats on the U.N. Security Council, that are guilty of institutionalised oppression and, of course, they are nuclear powers. How does the international community deal with an oppressive government that has the ability to destroy the world with nuclear weapons? It can only be by rational argument and the power of the persuasive word with the threat of total embargo. No armed action would ever be successful; so, it should never even be attempted. A nation that is isolated from the world by total embargo would soon want that to change.

Now, you see that international justice and the fight against oppression is, perhaps, the greatest challenge that mankind faces. Yes, our greatest challenge is to face the worst in ourselves at an international level. It could precipitate wars and, in all likelihood, given certain smaller players, it would. This could, of course, potentially lead to nuclear war. So the international community is faced with the difficult decision of whether or not to turn its face away from the oppression and continue to either be oppressed or allow others to be oppressed or face the possibility of nuclear war. As I said, justice is not easy. Total embargo itself is not easy, especially if the country has long borders. Isolation and peaceful entreaty through argument and persuasion, face-to-face at the diplomatic table, are, without doubt, the best way forward.

If governmental oppression of humanity can be eradicated, it would be the greatest victory mankind has ever made. It is far greater than walking on the moon and far greater and more important than discovering the Higgs boson. The reward for the eradication of human oppression is that our children and their children will enjoy a world of peace and justice and that is something that no generation of mankind has ever known. It is possible but it will take people with an immense power of persuasion and I for one, would employ philosophers as the diplomats and leaders for this task. Plato was right, again, in this case, that only a good philosopher would be able to lead a nation away from oppression and towards justice, whether that nation be his or her own or that of another.

Therefore, I implore the international community to discover and employ the best philosophers throughout the world and take them out of their classrooms and put them to work as diplomats working to fight oppression wherever it is found. Any philosopher that loves their field of expertise - and, by definition, they must - would relish the chance at using their power of argument to bring peace to the world and bring the world that much closer to a just and righteous place to live. After all, this is the goal of philosophy - not to teach philosophy, but to employ it to make an actual difference in the world. Philosophy and justice share a portion of truth and this common ground can be exploited to bring justice to the world for our children's sake and for their children's sake.

Justice - true justice - as I've said, is only possible by God and, so, you can count on the fact that God will help those who act in God's cause because God's cause is just. God's goals and reasoning is not like our reasoning; it is greater and far more-encompassing than we can imagine; but it is our philosophers rather than our scientists who have the best chance at uncovering those reasons because the philosophers understand the human condition within the 'scientific environment' in which they exist. This is why I say to employ philosophers as diplomats, as they offer humanity our best chance at discovering world peace through wise and peaceful negotiations. Philosophy is a far more intelligent way of changing the mind of oppressive regimes than by water-boarding - thus further oppressing - those who have been brainwashed to accept the reasoning of an oppressive regime. You catch more flies with sugar than salt, no? So, you will change more oppressive governments with strong philosophy than by covert military action.

So, people of the world, ensure your governments are acting justly. Then, those just governments can act together and utilise the philosophers of the world to alter the minds of oppressive governments through reasonable argumentation. This is a peaceful way forward but, yet, it is still not without risks. Risks are inherent factors to change and change is required. Not all change is progress but progress is impossible without change. All philosophers know this, as well as all logicians. Risks can be minimised when the 'opponent' (not 'enemy'!) knows that peace and justice is the goal, so, let there be no hiding of that aspect of the mission.

Oppressive governments will know, up front, that the goal of parley is peace and justice for all. There is no need for secrecy in this regard and any secrecy discovered will work against mankind. Spying is a dirty methodology when one can be honest instead; and honesty will be honoured more greatly. In a world where there is world peace, the CIA, MI-6, Mossad and other such organisations will not be required. In a world where there is world peace, trust can be found and these agencies can be turned into agencies that ensure trust, transparency and open relationships between governments. These agencies, when re-strategized, as it were - will be very useful in a world where there is world peace and world justice for mankind; they will simply have a different remit.

What about Sin and Damnation?

What is sin if there is only one actor in the system? That is, what is it that we humans should not do given the knowledge that there is, in fact, only one entity in the universe that truly exists and that that entity is God? In my opinion, wise King Solomon had the answer when he told us, in the book of Ecclesiastes, of the woes begotten of vanity: "Vanity of vanities; all is vanity." (Eccl. 1:2) Vanity itself is the root of all sin in a monistic system. But what do I mean by 'vanity'?

When an individual soul thinks "I", they separate themselves from the one that is. Vanity is when we think "I". This fundamental grasping of our own identity is completely counter to the concept of the oneness (that God is One without unity, as unity implies an assemblage of parts rather than an extension of the oneness) of God; thus the ultimate realisation in Hinduism that Atman (Self) is Brahman (God). In Ecclesiastes 1:9, Solomon says, "The thing that hath been, it is that which shall be; and that which is done is that which shall be done; and there is no new thing under the sun." The first clause of Eccl. 1:9 is another declaration of the oneness of God, saying that God, the thing that hath been, is the only thing existing and is, thus, that which shall be. The second clause pertains to the argument of fate vs. free will. In our space-time continuum, all events are extant in the whole of space-time. The future is just as much "there and then" as is the past. That which is (to be) done is that which shall be done. The grammatical concept of obligation inherent in the usage of the word "shall" is also relevant because God is obliged by Its very nature to perform every act at the right time and at the right place everywhere always. The third clause is saying that there is no new thing under the sun (a metaphor for God) because there can be nothing other than the one thing, which is God. It is also a metaphor for understanding that energy is neither created nor destroyed, only transformed from one form to another; because 'that which exists' is energy that has always existed and always will, there can be nothing 'new'.

In the Torah, eight of the Ten Commandments are negative commandments, i.e., those that prohibit behaviours. The first negative commandment is, "I AM the Lord thy God... thou shalt have no other gods before me". God states that it is a sin to acknowledge the existence of gods OTHER than Itself. It would be impossible for a monistic God to acknowledge an entity other than Itself, as 'monistic' implies that there is only one. After all, God is omniscient; It would know, beyond doubt, that there was no other. So, too, it is wrong and a vanity for man to acknowledge any other god but God. This is also in perfect keeping with the Islamic proclamation of faith, the Shahada, that 'there is no god worthy of worship but God' ("La ilaha ila Allah").

The second negative commandment is, "Thou shalt not make unto thee any graven image... of anything that is in Heaven or on the Earth... for I AM a jealous God." In this commandment, God gives Its reasoning for the prohibition. Jealous, in this usage, means 'demanding of complete loyalty'. One is not permitted to try to depict God as any one thing because God demands complete loyalty and, to be completely true to the concept of a monistic God one would have to depict the entirety of the space-time continuum in order to be comprehensive. Anything less is a vain attempt. To think that one could, in any item, truly depict God "in toto" is vain. It is also something that God cannot do. God cannot, in any creation, depict Itself fully except for the entirety of the Creation It has already created.

The third negative commandment is, "Thou shalt not take the Lord thy God's name in vain." Here, it's plainly stated. Again, to think that one could change one's destiny by calling out the name of God is blatantly vain. Plus, God helps those who help themselves. Remember that all events are extant in the whole of space-time and it is God that drives them all. There is nothing any of us can do to alter the will of God and to think we can is to be incredibly vain. Rather than calling out God's name in vain, we should re-double our efforts to affect our goals and know, firmly, that God will, naturally, always be there to help us.

The fourth negative commandment is, "Thou shalt not murder." To think that we are so powerful as to be able to snuff out life is vain. To an object of energy, all events can be boiled down to various transformations of energy. In our universe, we have discovered that energy is conserved and not lost. It only changes from one form to another. At the moment that we call death, there may well be a series of energy transformations such that the non-corporeal elements of our existence are separated from the corporeal but that does not mean that life, which is experienced through our consciousness, ends. Once a field of consciousness has been created, it is anchored to the Calabi-Yau space, which is outside of time. Because this field is predominately outside of time, it cannot cease to exist. It stretches towards time and a form of temporal existence when incarnated; but, its main place of existence, its home, in the Calabi-Yau space is atemporal. Energy transforms. That is all. Consciousness is, if anything, freed from the confinements of the body at death as much as it is when we dream. In our dreams, we can act without fear because there is nothing there, truly, but ourselves. To think otherwise is to deny one's own being. Life, in this case, more properly, one's ability to remain self-aware, cannot be ended so long as there is an extra-temporal aspect to the field. To think otherwise is vain because it denies the oneness and the continuity of God's self-awareness. Simply put, because we are immortal 'thinking things', our existence does not end and murder, as defined as bringing an end to another's existence, is impossible; so, to think that we have the power to do that is vain.

199

The fifth negative commandment is, "Thou shalt not commit adultery." This is about internal consistency, loyalty and acting in good faith. In creating this universe, God has said "These things will happen". And those things will happen. There is no changing the will of God. To act in such a way as to break our will - which is implied by adultery, as adultery is an act against a solemn vow to not act in that way - is to deny the oneness of God. There is no new thing under the sun. To think that we have acted in such a way as to break the will of God, is vain and it is, also, to believe that God is not omnipotent. If something happens, then it must have been in accordance with God's will, otherwise God is less than omnipotent and that is not the case. Also, we cannot act against our own will. Will is resolute. If it is one's will to do A, A will be done. If A is not done, it could only have been a desire to perform A, followed by a desire to not do A. Will is always performed. If adultery is perceived to occur, a wise individual should realise that it was never the partner's will to remain faithful but, rather a desire to remain faithful, followed by a desire to not be faithful. In another area of the Torah, Jews are entreated to not make vows lightly because vows are a declaration of will. If a man presumes to declare his will and acts otherwise, he soon loses the respect of his peers. Simply put, adultery is duplicitous behaviour and "One" cannot be duplicitous. That alone would be enough but God, also, acts ONLY on will, as God has no desires. Desires are based on a perceived lack of something and God lacks nothing. When you are all that there is, what could you possibly desire?

The sixth negative commandment is, "Thou shalt not steal." The well-respected Torah commentator Rashi states that this commandment pertains to kidnapping, i.e., the stealing of other people. In a broader sense, it deals with the concept of owning an individual. If you steal someone, you have taken them like you would take an object. You hold a claim of ownership over the individual. Slavery is a form of kidnapping where the victim is forced to work for the captor. To think that one is capable of

200

owning another is vain. To believe that we are powerful enough to own another human with a living soul is to deny the oneness of God by way of thinking that the Creation or at least certain aspects of the Creation are separate from God. What can you remove from God? All is all. Even the taking of items, which, at a scientific level, is simply moving an item from one place to another, does not remove them from God's presence and to think that you can steal, to remove an item or a human being from the presence of an omnipresent God, is vain.

The seventh negative commandment is, "Thou shalt not bear false witness." To think that you can hide the truth from God, who is omniscient, is vain. To believe that we are actually capable of preventing the truth from being known by God is, given the premiss of an omniscient God, a logical tautology. We know the truth, therefore, the truth is known and, through us, God knows the truth. Our thoughts aren't really ours; they are the thoughts of God that have been loaned to us for our use while we are incarnate. Our recognition that "others" may not know the truth does not prevent the truth from being known by God through ourselves and it is completely vain to think otherwise.

The eighth negative commandment is, "Thou shalt not covet anything that is thy neighbour's." Buddhists understand this completely, which is why their faith concludes that it is hurtful to oneself to have attachments to material things and material things include humans because of the material aspect of our temporal existence. To desire material things is to become attached to transient objects that may be destroyed or lost at any time; it is a very risky venture. To feel that God has not provided us with all that we need is vain. We will encounter, in our lives, everything that we will encounter. There is no part of our experiences where we can gather more to us than was allotted to us. To think otherwise is to deny the oneness of God and God's Creation. Our lives are extensions of God's existence and there is nothing that is not already, completely, God's.

These commandments express actions that cannot be performed by the One God:

1) The One God cannot recognise another God. (You should have no other God.)
2) The One God cannot create any item that is a subset of the whole that can fairly represent the whole. (You should not make any image attempting to depict God.)
3) The One God cannot change that which will be; rather, God makes it that which will be. (Both taking the Lord's Name in vain and committing adultery are derived from this principle, as you should not ask God for help when YOU can help yourself nor should you act duplicitously.)
4) The One God cannot extinguish at any time that which has a portion outside of time. (You should not murder)
5) The One God cannot remove anything from the whole of space-time. (You should not steal)
6) The One God cannot NOT know the truth for it is omniscient. (You should not lie)
7) The One God cannot add anything to the whole of space-time. (You should not covet)

These negative commandments tell us that we should not act in a way that denies the oneness of God and imagining that we have the power to actually do these things is the vanity of all vanities to which Solomon referred when he said, "Vanity of vanities; all is vanity". The more absorbed in the physical we get, the less absorbed we are by the One. Roman Catholicism has contributed greatly to our understanding of the spiritual harm done by indulging ourselves by outlining the Seven Deadly Sins: Lust, Avarice, Vanity, Pride, Sloth, Anger and Gluttony; if you have trouble remembering them, try the monistic mnemonic: Look, Acting Vainly Pits Self Against God. These desires prey on our sense of self and make us act ungodly, as God has no desires; so, we, too, should act selflessly. I will restate that God has no desires because God already has everything and it is illogical that God would want anything or desire anything because God knows that there is nothing in existence that is not already in God's possession. We only have a sense of self because God has a sense of self and lent that sense to us. Because that self is the selfsame as our self, we literally owe our selves to God. This is the realisation that Atman is Brahman and is the realisation made by the 'Ipsissimus' magician, as 'Ipsissimus' is the Latin for 'selfsame'. And, for that reason, all texts that claim to be revealed by God decree codes of behaviour that are in keeping with an objective of promoting selfless behaviour.

In the New Testament, Jesus is asked regarding the greatest commandments of the Torah. In other words, what is the greatest bit of advice in the Torah that would help mankind become closer to God? His response was, "Love the Lord your God with all your heart, with all your soul and with all your mind"; i.e. think God, feel God, be God. And Jesus followed that by saying the second greatest commandment is, "Love your neighbour as yourself". The Hebrew preposition 'K' translated as "as" does not mean "as much as" but, rather, "as if equal to" like the "as" in "A is to B as Y is to Z". Jesus is teaching that we should love our fellow humans "as if they were" ourselves. And, of course, love is a kind of spiritual gravity, as it is the attractive spiritual force. We are entreated to fully attract ourselves to God, gravitate towards God, and to consider others as no different from ourselves.

Just to be complete with respect to the 10 Commandments, I wanted to express what I feel is the reasoning for the existence of the two positive commandments. These two commandments fulfil a very unique niche and they are:
1) Honour thy father and thy mother
2) Remember the Sabbath and keep it Holy
The first, honouring one's parents, if everyone did throughout all time, would establish a firm link to one's past. If a child honours his/her parents - if, indeed, they know who they are - they establish

a link with them and that forms a link to their past. It gives us a sense of personal history and helps us to understand what led up to our existence. The more generations that take part in this process, the greater an individual's personal history is known and the more chance a child has to understand the events that led to their existence. Honour also involves love, trust and patience and many other virtues that engender a person towards another. All of these enable an individual to have a greater sense of self and self-respect as well as respecting those that came before them. This honour and respect towards one's ancestors is a vital aspect of Shinto and several other Oriental faiths. This commandment, simply put, forms a link to the past and having a strong link to the past is invaluable.

The second positive commandment is the link to the future. It isn't about remembering that we should rest one day of the week or remember which day is our day of rest. It's about having something to which we look forward. The value of knowing that, for example, 'next Saturday (or Sunday, if you're a Protestant Christian) I can have a full day of rest', can come as a great relief and act as a stimulus to get us past the hard work-days of the week. Keeping that day holy is the aspect that makes it important and puts pressure on us that we not forget it. Put together, the concept of remembering the Sabbath and keeping it Holy formulates a way that, each week, continues to drive us forward throughout our lives towards the future; but it gives us hope, as we look forward to that next respite.

Thus, the two positive commandments form our links to the past and to the future. They help establish bonds that aide us in retaining the integrity of our family units and they give us a perfect reason for never giving up by maintaining a continual hope for the future. So long as we do these two things now, in the present, we reaffirm and re-establish our links to both the past and the future. There is great subtlety to these commandments and it is that kind of subtlety that, once again, is exactly the kind of thing one would expect to come from a God who has perfect wisdom. Part of that subtlety is in the fact that honouring our parents is a selfless act and, so too, is having hope for the future. They do, in that respect, aim toward the same goals that the negative commandments do, albeit by different means.

So, to summarise, in a monistic system, the focus is the Self. The aim is to realise that one's own self is no different to the One Self of God. By us acting selfishly, we increase the differentiation between us and God, so vanity or selfishness is sinful. Equally, by us acting selflessly, the monistic virtue, and putting God first in our hearts our minds and our very being, we lose our transient selves and gain our True Self.

That covers sin, but what of damnation? To understand damnation is to understand divine punishment. So, if the crime against self is self-inflicted, so, too, might be the punishment. Remember the words of the Torah, "Love thy neighbour as thyself." It leads quite naturally into the application Jesus taught regarding doing unto another as you would have them do unto you. Combine the concepts and you have something like, "treat others as yourself." Since there is only one, how we have treated others is exactly how we have treated ourselves; for example, if we take our left foot and use it to kick our right shin, we use two 'separate parts' to inflict an injury, yet, if we look further up, we see that the two 'parts' are actually extensions of a single, greater body. I think it is this simple fact that will be explained to those who have died after their death, at their 'judgement', if they have not discovered it to be true before then. When we fully realise the amount of thoughtless injury and disregard we have sown in the world in our short lives, most of us would be appalled. This horrific revelation coming to the unsuspecting soul could well be enough to send it into a self-directed angry insurrection, hell-bent, if you will, on revenge. And it's worse; we must have, because possibility and probability demand that it exists, situations where God's judgement will stand against a soul even if the soul is remorseful.

This is, no doubt, for extreme circumstances when some recompense must be paid in order to account for the required 'equal and opposite reaction' that is requisite of physics. However, as it is also plausible, there must be situations, at judgement, where the individual soul is so overwhelmed by his/her actions that they damn themselves in order to pay recompense - they realise the punishment is deserved. In such a case, the soul/mind, now free to create like in a dream, creates a suitable punishment for itself in Hell or other suitable underworld/afterlife mechanism, based on concepts encountered during their incarnation, and plays out its punishment for however long it wants to play. Note that this is all done by a mind that, while living, had not realised its union with the One.

A fully realised soul, though, knows that it was God who acted through them. All things occur because it is God's will. This is a hard truth for many people as it is common for us to perceive evil in the world and how could an omnibenificent God permit evil? An omnibenificent God can't but God is not omnibenificent. God created Satan/Iblis in order to allow us to perceive evil; thus, God is, ultimately, the source of evil. Rather than being omnibenificent, God is the 'Most Merciful' and 'Most Just' thing that there is and there is an enormous difference between those concepts. Equal and opposite reaction demand give and take for actions made. The problem is, again, ours. We perceive evil because we see events that produce in us feelings we don't like to have. Evil is usually perceived in acts that are considered to be non-life-affirming and/or restrictive of civil liberty. So, physical assaults, murders and wars are seen to be evil because they are non-life-affirming acts and rape, theft and coercion are considered evil because they restrict civil liberties.

But it is how we react to perceived evil that shows us for what we are. Our reactions to our perceptions are vital because we have that split-second (or, usually, much longer) to consider how we respond to stimuli. It is that moment of consideration that we have that places the responsibility for our actions completely back on us and not God. This is part and parcel of the perception of free will; we have time to speculate about what may be, and that makes us responsible and gives us an undeniable sense of free will, because we cannot see into the future to know what we will do. So, our reactions are what matters and each of our reactions is, in fact, an action itself, which requires an equal and opposite reaction. In order to weigh us fairly, it requires God's knowledge of all our intentions; and this is how and why some judgement must be made. Considering all this it is vital to each and every one of us to ensure that all our actions and reactions are the very best we can make and those that will, God willing, lead the whole world to a better place - and that is an enormous responsibility that has been thrust upon us. Should we be glad or incensed at having that responsibility thrust upon us? Be glad; because, if you are incensed, you're one of those happy few who are alive enough to change things so that you are glad.

Punishment is usually doled out by an authority upon an individual for perceived wrongdoing. In our day-to-day affairs, our civil authorities punish criminals for acting in a way that is seen by the authorities to be wrong. Mankind punishes its criminals in many ways from fines and personal sanctions like points against one's driver's license to confinement in prison and death. But, when the authority is God, the situation is radically different. The authority is omniscient and, in this case, the authority is also the creator of the accused. The Creator, being omniscient, would have known long before the accused was born that he or she would make the choices they would, as a result of being omniscient. So, it is a logical conclusion to assume that, if God actually imposes eternal damnation on Its creations, It does so by creating individuals who are, essentially, damned from the moment of their birth. Some would argue that the accused, while alive, apprehended free will and didn't have to make those choices that would lead to his or her damnation. But I would argue that that is

207

immaterial due to God's omniscience; God knows that he or she will make those choices, so, from God's perspective, the individual is damned from the start. Now I ask you, if God is supposed to be 'Most Merciful', what divine purpose is served by creating throw-away souls? It's tantamount to having children and, the first time you get angry with them, you decide to boil them in oil. If God has a shred of mercy - and God must be 'Most Merciful' - God would not act like a spoiled brat and eternally damn those who act in a way in which It knew they would. I can't see the logic in a God who acts less humanely than your average human. The concept of divine damnation defined by service in Hell for eternity is not in keeping with the way an omniscient and all-wise entity would act. In fact, it's a ridiculous assertion even given a dualist scenario, because it purports that an all-knowing and all-wise God remorselessly creates entities bound for Hell with no real hope of the salvation presumed to be available to everyone. That kind of action is not 'Most Merciful'. So, there would always be a limit of time spent in Hell; otherwise, no lesson can be learned and no gain for the individual made when there, otherwise, could have been. This is also backed up by a Qudsi Hadith, a statement made by God through the angel Gabriel to the Prophet Mohammed (saw) but not intended to be a part of the Qur'an, which states that "God's mercy prevails over His wrath". If this is true - and it must be for logical reasons - then there is no requirement for a true eternal punishment. This is because one must remember that any space-time time that passes while your soul or consciousness is outside of space-time will **seem** like an eternity. Any time spent in a space that is not linked with time will be experienced as eternal; so, whilst punishments may seem to be eternal and feel to the individual as if they are, God's divine mercy prevents them from actually being eternal. God's wisdom has created a geometry for space and time that allows Its mercy to prevail over Its wrath and this means that no soul, with respect to time as experienced in space-time, will spend an eternity in Hell. However, all time spent outside space-time will seem eternal and so punishments will feel eternal - nevertheless God's mercy must prevail.

Could the same be true regarding rewards in heaven? Well, the flip-side to the Heaven-and-Hell coin is a completely different story. Given a Most Merciful God, rewards in Heaven would, most likely, remain eternal, as surely, an eternal reward is more merciful than a reward that is less than eternal. So, rewards are forever but punishments will only seem like they are. It's a subtle difference and one the individual soul cannot discern; but, that difference is a vital part of God's attribute of being 'Most Merciful'.

Some, like St. Anselm, may see this leaves a hole in the logic in that a God who can and does perform eternal damnation is greater than one who can and does not; to this there is a simple answer (but, to be perfectly honest, it took me a while to work it out). Eternal damnation is reserved for the one created entity that truly deserves it, the entity that tempts all the others away from God's guidance. Iblis, Satan - call it what you like - will be the one to have eternal damnation as its punishment. When all the universes are completed and Iblis' respite has, finally, come to an end, then God will deport that entity to Hell forever. This was the terrible bargain Iblis made by not bowing to Adam when God commanded it to do so and asking for respite. So, the 'trick' that was played out on Iblis - that showed Iblis as so loyal to God that it would not bow to anything less than God - is the single act that grants God the opportunity to damn one soul forever and, in so doing, retain omnipotence and still maintain the attribute of being 'Most Merciful', as God has only made one eternal punishment. Iblis' loyalty is also the determining factor that forces Iblis to accept that fate as completely fair and just, which is why Iblis works so hard at tempting mankind - in order to deserve that punishment. God leaves no logical holes and has foresight beyond that of humans; yet, sometimes, we are granted a tiny peek inside what Sir Tim Rice, in *Jesus Christ Superstar*, called God's "omnipresent brain".

Is Resurrection Possible?

Yes, and it may be far easier than you think it is. Think about the first time you rode a bike. It was easier the second time, wasn't it? So, too, with resurrection, the re-creation of each human is easier than our original creation and, yet, here we are. The fact that we are here, now, is far less likely than any secondary appearance we make later. The first time of doing anything is more difficult than repeating a process, as you already have a successful attempt behind you. Not that God requires practice but, even for an omnipotent entity, doing something a second time, whatever that something is, is easier than doing it the first time. This is just simple logic. And, God retains our blueprint in the 'clear record' that is the space-time continuum.

In the case of human resurrection, our current existence gives God a template from which It can borrow in a future time. To use a computer analogy, resurrection could be done in a way not too dissimilar from what we call 'copy-and-paste', although I think it is more likely that the methodology would be more on the lines of utilising the original instance as a blueprint for re-arranging matter in a similar way in a future time, which is a re-creation that would be easy for a Universal CPU that contains our blueprint as evidenced by our current existence. Perhaps the analogy is best expressed by a methodology akin to the Star Trek process of 'Transportation', where the molecules are re-assembled to form the entity in a different place/time.

Science does not speak about resurrection, so I cannot give a current scientific view of it; however, my physics most certainly allows for it. It does so in both the 'copy-and-paste' method, which is afforded due to the fact that the space-time continuum always contains all spatio-temporal events and, therefore, always contains a perfect record of all that we have done and it allows for it in the 'Transportation' method by a re-arrangement of molecules into any given individual's previous form. Matter can be transformed into you, see?

It is my belief that the structure that forms the minute skeletal structures within our nervous system that is formed by that tubulin framework is what forms the interface between our physical bodies and our spiritual fields of consciousness. Thus, when the body is re-created each cell is recreated and the link between that body and the field of consciousness that it contained is re-called, per se, and this is the essence of resurrection. At least this is the methodology I suspect would be used in the 'Resurrection of the Dead' as referred to in Judaism, Christianity and Islam, which happens just prior to the end of a given Big-Bang-to-Annihilation cycle. It's what serves as a final proof of God's existence to believers and non-believers alike.

Even identical twins have ever-so-slight mutations that make their DNA not match perfectly. This is the reason why, when we are familiar with identical twins, we can tell them apart. They are not perfectly identical and, so their fields of consciousness are not perfectly identical and each has its own unique identity. This holds true for identical triplets and quadruplets and any other number of so-called identical offspring. The truth is that they are not perfectly identical and the difference is what allows them to link to a unique field of consciousness. In this way, God has made us all unique. All it takes is one tiny difference in the coding and the tubulin arrangement is altered offering an interface to a unique field of consciousness. So, don't worry if you are one of identical siblings; you will not be confused in the resurrection.

As for why a resurrection is predicted, we can only turn to scriptures for that. I prefer to use the Qur'an as a guide for this, as it is the only revelation about which we can be completely sure that it is, word-for-word, the same as when it was given. I do not mean to imply by that the other scriptures are without merit when

212

they discuss resurrection - this is not the case - but I prefer to rely on what is most reliable. Resurrection is to precede the end of the world so that all souls - fields of consciousness - can know that God is real and that God has the power to return them to their previous forms and, then, destroy the entire universe in order to create a new one afresh. It is done to prove equally to disbelievers that they were wrong and to prove to believers that they were correct.

Also, if the end of the world comes, as I believe it will, in matter/antimatter annihilation, it will not be a painful way to die. The time it will take to turn any human body into light during such an event would not give the nervous system time to process any pain before the process was finished, so, it will be painless - but death itself is always painless. What will be painful is for the unbelievers to experience the fear and dread of any judgement that comes afterwards. For believers, some will see the panic around them and even some, themselves, will panic and will experience emotional pain. If you are a believer I would think you have less to fear by this event. If you have led a decent life and done more good than bad, then God's mercy will see you through. Trust the one who is Most-Merciful.

Resurrection, in a sense, will be done to prove to those who 'choose' to not believe, that they are mistaken and they will have much to fear. Unfortunately, it is altogether likely that it will be impossible for an atheist to convert once resurrection has occurred. It is certainly impossible for those who have died and are re-created at the resurrection to convert and it would be unfair to allow those who are living at the time of the resurrection to convert, as the simple fact that they have not yet died will not benefit them when there is less than 8 minutes of time left.

I say 8 minutes because the sun is roughly 8 light minutes away from the Earth. The first sign of the annihilation, according to the Qur'an, will be seeing the sun annihilated and, once we have seen it, we know that it happened 8 minutes ago. The next thing to happen is the annihilation of the Moon, then the Earth, this would happen in a matter of seconds once we've seen the Sun splat against the backdrop of the wall of anti-matter and turn into red light. Some on Earth, from certain vantage points, will see a very odd thing: it would appear as though the sun were rising in the West as the red glow of the annihilating Sun would be cast in the Western sky; thus, the Qur'an's statement that the Sun will rise in the West.

When the annihilation reaches the Earth, the atmosphere on the near-side would vaporise and it would appear as though the sky were dissolving. This, in truth, is exactly what would be happening. Then, if you were near mountains, the tops of them would appear to turn to powder and light and then the Earth itself would vaporise with all men and animals on it. To those on the opposite side, it would appear as though the Earth were flattening out against the backdrop of the wall of antimatter, thus the sea, if you were near the coast, would look like it was no longer curving with the roundness of the planet but would appear to stretch out further before disappearing altogether. This, of course, matches exactly with what the Qur'an says and it is in perfect keeping with the physics I set forth above.

It will be both a horrendous and a glorious sight to behold and, because the resurrection will precede it, we all get to see it. The resurrection then, as it happens prior to the annihilation, will be the clue as to what will come shortly thereafter. Watch, therefore, for the resurrection and, when that occurs, you know you only have a few short minutes to live. If you're a believer, gather your family together and spend time praying, as that would be the best way to go. If you are not in a position to do that, then try to calm those around you, if any, and pray that your good behaviour in the face of certain death will stand in good stead on your behalf - even if you are an atheist! Remember that God's mercy is greater than any of us can imagine and, although it may be that an atheist cannot convert in those last few minutes, they can show, at that point, that they accept the truth and, if they act in a beneficial way towards their fellow humans in the face of certain annihilation, that can only help them.

Also, it may be that those who have died will be re-created near to their loved ones. It is equally plausible that they will be re-created near their own graves or wherever they last existed. I do not wish to raise the hopes of people but, if our loved ones are resurrected near to us, it could be a great comfort to us and to them. It seems to me that, if God is Most-Merciful - and God IS Most-Merciful - then God would do this for them and for us. But, as I said, I don't want to raise false hopes. There is little doubt that anyone on the planet would not notice the resurrection when it occurs, as there are so many places where the dead have fallen or been buried that it would be very difficult to be someplace that is truly away from it all; unless you happen to be exploring Antarctica or the Atacama Desert. Therefore, you will, in all likelihood, be very aware when it happens, if it happens during your lifetime. That said it is also likely that some people will be asleep. I would think, though, that, given the current population of the Earth, there will be someone awake in almost every neighbourhood and the turmoil and panic that will ensue because of the resurrection would soon wake those who were asleep. I truly believe that, when the time comes, we will all be awake and know exactly what we are facing.

That is why I implore believers to pray. For Muslims, if they can execute one raka between the resurrection and the annihilation (Al-Qiyama), it would be in their best interest to do so. After the raka, stand, as standing is the essence and meaning of 'Al-Qiyama' and proclaim that God is Greater (Allahu Akbar). If a full raka is not possible, then say the Shahada and the Fatihah, if that is not possible, then just the Shahada and then stand. For Christians, saying the Lord's Prayer would be most helpful, and then stand to be accounted. For Jews, there is no doubt that it would be best to re-affirm your faith by saying the Shema as many times as possible; luckily, for Jews, the Shema is rather short (Shema O-Yisrael, Adonai elohainu Adonai achad. Meaning: Hear,

215

O Israel, The Lord our God, The Lord is One), then stand to be accounted.   For other faiths, state your affirmation of faith whatever that may be and stand to be accounted.   To die with these prayers or affirmations on your lips will count heavily in your favour.   For atheists, as I said, just help your fellow humans and accept the truth that you were wrong; and, perhaps say, 'God help me'.

I know all this sounds rather dire, but the resurrection is said to precede the end of the world. If that is true - and there is no evidence either scientific or scriptural to deny it - then the situation is dire. It is better to face the truth even when the truth is dire. We don't know when it will happen but we are told in Scripture that it will and, given my model of physics, it will happen. So, we'd better include the concept of 'how to act just before the end of the world' in our Religious Studies, then. Besides, you can't claim it's unfair, as the entire world is destroyed; no one gets to avoid it. It doesn't come any fairer than that. The annihilation happens to the entire universe, so even extra-terrestrial life forms would face it, too. And, all that energy is wrapped back around to begin a new universe with a new Big Bang and a whole new sequence of universal events. All energy is conserved; the physical energy is re-used to re-create the new universe and the spiritual energy is conserved as it is recalled back to the Calabi-Yau Consciousness Space, which is its home, to be judged perfectly fairly by the Most-Merciful entity that exists.   Nothing is ever lost... ever; how can God lose God?

So How Should We Act?

It all boils down to a philosophical approach. Each of us is here to do that which we will do. So what is it that you would like to do? Remember the old adage of "God helps those who help themselves"? God works through Its creation. Armed with the rather dichotomous and Rumsfeldian knowledge that we know we don't know our own future, irrespective of how fixed it is, you still have to get there and you will get there. So we should each be trying to be the very best 'us' we can be. Also realise that, in a monistic system, the only will is that of the One. We can sit back and be depressed and think, "Why bother, if my life is fixed?" or we can take a more positive approach and think, "This allows us to work towards doing the things we've always wanted to see done in the world!" Actions cause reactions and that is an overriding factor. This is a universe where action is all important: it's what gives energy momentum and DOES. So if we want to do something or see something done, we must strive for it. The knowledge that when we act, it is, in fact, God acting, should serve to infuse us with great confidence knowing that each and every one of us is an Ambassador of God's Will. Of course, some people who put themselves before the One could use that to justify horrifying actions, most people would prefer to be of some benefit to the world and would naturally act ethically.

Jesus reiterated the importance of the Torah's commandments that we should put God first and love one another as if there were no difference between us. With those incredibly monistic guidelines, if we but followed them, we would act more empathetically towards one another and, if we tried to understand others more, we'd probably learn a lot more about ourselves in the process. Imagine a world of people that really cared for one

217

another. It can only happen, though, one at a time; and, as there is only One, one can only motivate oneself. This last statement, "that one can only motivate oneself", is a key realisation in understanding the knowledge that we are an extension of that One and know to not separate ourselves from that One as, indeed, it serves as the basis for true empathy.

It's time to move out of the Stone Age and into the Atone Age. In the Atone Age, we are 'at one' with another. And atonement comes as a result, an effect, of forgiveness. When we forgive another, we are reconciled, i.e., in atonement, with them. We are, no longer divided; but, rather, at one with one another. Now, this may sound a hard teaching, but, when this logic is applied to its greatest extent, i.e. the realisation that we are all One, we must acknowledge that, because WE are that One, we accept the sins of the world as ours. By the sheer fact that we are, in a larger sense, one with one another, another's 'sin' is ours, for there is only One. It was this outward teaching without the inner knowledge that was distorted by Paul into Jesus' vicarious sacrifice.

Jesus knew that what was true for him was true for everyone else, for there is only One. But, over time, the message that anyone and everyone should understand that level of continuity between humans and consciousness was lost in Christianity - with the possible exception of the Gnostics. Unfortunately, due to Pauline teachings, it became a one-man show where only Jesus could do that; and, in Christian/Pauline fact, has done that and we need only recognise Jesus' acceptance. But his concept of taking on the sins of the world was a result of realising that we were all part of a greater One and that, if any of us are perceived to sin, then we have all sinned until that sin is forgiven. So he taught that we should, above all, love God, i.e. the One, with all our heart, all our might and all our being. And he taught that we should 'love your neighbour AS yourself'. And if we loved one another - that is - if we treated one another as if we were all intimately related at the level of our consciousness itself - as well as through DNA - we would act in such a way as to more easily forgive one another; we would, naturally, be more charitable. We are related in just such a way; we are all, at least, distant cousins of one another. Furthermore, Jesus taught that we should 'love our enemies'. In other words, live your lives as reconciled with others as you can possibly be; for if you did that, you would have no enemies - nor would anyone, if each of us acted in such a way. World peace is only a tiny step away once all thoughts of 'enemies' have been banished from our consciousness. I know, I ask too much but I'm an idealist as a result of being a philosopher; but, without ideals, we have no target towards which we should aim.

If consciousness does exist outside space-time as I purport and is, therefore, eternal, and if one accepts the sins of the world themselves, and then, forgives themselves for those sins while in that 'eternal' state, such an individual has forgiven all sins. And, as we have learned, forgiveness has the potential to change the past. This leaves us with a situation in which Jesus apparently died on the cross, yet lives, because his faith in the oneness and his acceptance of all sins followed by his forgiveness of himself while in the tomb (i.e. in that state just after death, when consciousness is, again, free and eternal) - the One Self, altered the past. It was as if he had disappeared and reappeared. Well, that is exactly what would happen if, at a point outside of space-time, the consciousness of the One was changed in that the alleged offences that Jesus had committed - not just the incident in the Temple and the 'kingship' implied by claiming to be the Messiah that angered the Priests, but those that he accepted upon himself as a result of his realisation of the oneness of humanity - had been forgiven, the Priests, thinking him dead, had no argument with him, and he was reconciled in himself; his 'offences' were gone, and those that had been held against him specifically - those, like the incident in the Temple and certain misunderstood teachings that led the Priests to

219

want him put to death - were also gone, therefore, no reason for Jesus to be dead. So he re-appeared.

Granted, this is a vast re-interpretation; but, if this is true, then if we all follow that example and can take the sins of the world upon us and realise that we are, in fact, at one with another, then perhaps, this is the mechanism behind the resurrection of the dead. The concept of 'taking up the cross' is this concept of each of us taking upon us the sins of the world, as Jesus purportedly did according to Christianity, and then, forgiving ourselves those very sins; but doing so out of an understanding of the Oneness of God. It's a hard teaching, but, if forgiveness really can alter the past, then this kind of conscious realisation across all humanity could have huge results on the past! And, if it doesn't bring back the dead, it can't do any harm by trying it, can it? The worst that can happen is that all those alive are forgiven by virtue of each of us forgiving one another and we can enter a time where all humanity realises that God or the 'One Thing that Exists' is what rules, and we will have, at least, entered the 'Kingdom of God' on Earth by knowing, through physics, that we are part of a greater One and that it is that One that rules. If we do that, then this is the beginning of the 'Atone Age'. But I'm an idealist.

Why Should We Act This Way?

It is my sincerest wish that this book will aid many on their individual searches for the truth about existence and what our role is in the universe. It is my greatest, most grandiose hope that this book will serve as the catalyst that will raise human consciousness to the next higher level. One in which we recognise and know, beyond all doubt, that we are part of a greater plan and that each one of us is vital to the whole of that plan. We must, as individuals, completely and willingly, want to learn about one another and help one another because we will understand in our hearts that we are all in the same boat - the boat we call Earth.

We must put away ancient and less ancient hatreds and move past the artificial barriers of race, gender, age, sexuality and, of course, religion and anything else that we rely upon to separate ourselves from one another. We are one species. Only through a unity of spirit can we make it past the next few challenging decades. It is my belief that, in these next 4 decades, many of the old eschatological signs of the 'end times' will be played out across the world of modern politics. I am not trying to be a scare-monger; rather, I am trying to set out a warning that there are people working to make those ancient prophecies come true. And they are so deeply entrenched in the inner circles of worldwide politics that they have the resources and the means and the motivations to make their dreams of Armageddon a reality.

Of course, this makes me sound like a mad man - just another oddball conspiracy theorist - think what you want to think. But I have done research that has shown me that this is, in sad fact, the case. I cannot reveal or unveil the real conspirators, as that would be, simply put, suicidal, because it would make me an obvious target of theirs; and I am not suicidal. Most of us suspect that our governments are involved in many secret operations and they are. Most of them, though, are based in economics: how to get more oil, how to control the modern oilfields and the like. These are obvious. But the way they go about it stirs up feelings of hatred among people that, before, had no hatred of one another. This is the subtlety of beginning Armageddon. Hatred is their key and it is that hatred that I want to pull the plug on. If people can stop hating one another and realise that the main reason for their hatred of others is that their governments are feeding them disinformation about their so-called enemies, then this slide towards the end can be averted.

I originally wanted to have this book published on 21st December, 2012 in order to take advantage of the marketing aspects of that day - the end of the Mayan Long Count Calendar. One of two things is expected from that day, either the end of the world or some new piece of information that would bring clarity of vision to humanity and allow us to step into that 'New Age'. I want this book to represent that latter aspect. I do not believe, for a moment, that the world will end on that day nor do I believe the Mayans did. Rather, the end of the Long Count is like December 31st is in the Gregorian calendar, the next thing that happens is January 1st, and the cycle repeats itself. It's just a calendar, which is exactly why it just describes a cycle. However, considering that there are certain people trying to bring about the 'end times' in a variety of ways and that the world is ripe, in certain ways, for such a cataclysm, I decided to bring the date of publication forward in an attempt to give humanity a head start on defending themselves from these peoples' agendas. Time is of the essence.

After the ravages of World War I - the Great War and the so-called 'War to End All Wars' - was World War II, which made World War I look like what it was, a rehearsal for the real war. Since World War II, every war seems to have its own built-in holocaust. World War II had the Jews being singled out. Well, they were only first on the list, the Gypsies and homosexuals were next followed closely by anyone who had more than a certain amount of melanin in their skin - truly the lamest of reasons. The Cold War came with all the promise of a nuclear holocaust; but that was brought to an end by the best role Ronald Reagan ever played - that of the President of the United States - and the miraculous 'Star Wars' defence system. It drove the Soviets to spend all their money and go bust. The Chinese were laughing and learning. In the midst of all that, though, was Vietnam. To the U.S., it was a 'Police Action' and not a war. But the sad truth was that this police action gave the Chinese the benefits of using Southeast Asia as a test ground for belligerence. The Killing Fields was the holocaust of that war: two million dead, a quarter of the population of Cambodia killed by its own leaders. Then there were the atrocities that occurred in the Plain of Jars. How did we, humanity, let those things happen after World War II?

Communism is, by its own admission, atheistic. Yet it demands that people believe in it. This is a terrible dichotomy for those who both demand belief but also require people to not believe. Perhaps this is one reason why it has never worked as Marx hoped it would. The Chinese now, the country with the largest human resources available, are still communists; but, now they've finally understood how to use capitalism. They discovered that, in truth, capitalism is ruthless, so, automatically, it appeals to the atheist, who has no God to fear and 'knows' that there will be no retribution. Their latest trick is to slowly buy America by buying its Treasury Notes and Bonds. This will, within the next decade, give them the power to dictate American policy safely from their Far Eastern shores. How do they afford it? They rape and ravage the natural resources of Africa and sell the commodities. I suspect that, for this paragraph alone, this book will be banned in China; but I still might be able to strike a bargain with the Chinese Government - they are not unreasonable, that is to say they have their reasons.

As I've said, communism is, by nature, atheistic and, therefore, feels no guilt and has no conscience about how it works. Individual atheists are, usually, no great worry. In fact, they can be some of the kindest and most ethical people you can meet. But institutional atheism - atheism almost practiced as a religion - is, perhaps, the most dangerous religion mankind has ever created. It has its scriptures in *The Communist Manifesto*, Mao's *Little Red Book* and, more recently, Prof. Richard Dawkins' *The God Delusion*; but it is insidious in that it has no feeling of recompense. The true atheist believer is absolutely sure that they can act in any way and never have to pay the piper, as it were. But the truth is very different. They are wrong and there is a piper that requires payment. This piper is, of course, God. God is using these people, ultimately, to give humans one last example of a very good 'bad example'. Apparently, we haven't learned from our prior mistakes, as we still have our petty hatreds. The Middle East is a classic example and one that serves to endanger the entire world.

There is another player in this 'Road to Armageddon' that is kept well hidden in the midst. This player is a collective of, essentially, Satanists, for lack of a better term. They are rebels against God who will ally themselves with the institutional atheists and international governments for a common goal: wiping out the believers in God. And that is the essence of the impending Armageddon. It is a secret war that works by stirring up the hatred and distrust between those who believe in God, yet who believe differently. Those differences will be exploited and will be used to stir the pot of strife. They will turn brother against brother, monk against monk, in order to sow the seeds of hatred and, if they are successful, they will have believers kill one another until all the believers are gone. Any remainders who might have the thought that they have 'won' by surviving would be slaughtered in the end by the institutional atheist/Satanist front and the world will be a Godless place. This is their goal and they are well on their way. They must be stopped.

224

I've alluded to the next forty years as being crucial; this is, roughly, two generations. We are at the crossroads at this point in time. We can raise our children with the truths I present and they will raise their children similarly and this will prevent the problem. The alternative is that we ignore the truths laid out here and sit on our hands and wait to be destroyed, slowly but surely. Godlessness is on the rise. Magog, in the form of the current Chinese regime, is growing stronger. Gog, the automaton that may well be a Chinese-controlled America, is growing in the womb. Could anyone resist these two powers joined together? Remember, America's own constitution requires that the State has no and promotes no religion. This can be used against it... and I believe that it will be at some point in the future.

The individual who will be deployed as The Anti-Christ - Ad-Dajjal in Islam - is, I believe, already alive. He is still just a baby, but, nevertheless, this entity has been created and will be groomed for his future, which is to play a false role of the Saviour of mankind. This is why I, in my 'Advice to Religions' section in the *Practical and Social Applications* part, give instructions on how to thwart this by doing that which will bring the real saviour. I don't believe that the ancient prophecies were lies; rather, they were great insights into the future of mankind. That once far distant future is now our very near future. Jews and Christians and Muslims can work together; they really can. I've seen it done at various workplaces. Why not do on a larger scale for the greatest cause of all: the survival of true faith?

Funnily enough, at this very moment, while I write these particular lines, my house is infested with clothes-eating moths. And by infested, I mean the top bedroom has about 200 in it at various stages of growth. The symbolism is very poignant. The secretive, destructive force that flies in the night, it eats away at our outer garments and does so without consideration. I see it as a symbol of blind destructiveness that waits until the time is right to move, yet is stupid enough to fly into a flame. It reminds me of the

225

line in Job 13:28, "*So man wastes away like something rotten, like a garment eaten by moths.*" It's not so far off the truth of mankind's perilous situation.

I implore you to read this work with the thought in mind that mankind is threatened. It is threatened by elements of itself that would have you believe that there is no God and that only Science is the way, the truth and the life. Science is a way towards understanding scientific truth, yes, but it is, by no means, the only way of understanding truth nor is it 'the life' in any way. Science cannot give a reason for consciousness other than, "that's what the brain handles." We are far more than the sum of our parts and we must learn to lean on one another and not mind that others need to lean on us. We are all in this together. We are one. Let us not forget our humanity and be led down a garden path only to be bludgeoned by a Godless robot that wants to cleanse our minds (read 'brainwash') of our silly and quaint beliefs.

True science does not have a godless agenda as Einstein said, "*Religion without science is lame and science without religion is blind.*" Let this work act as a bulwark against the slow-but-sure creeping of the New Atheism. It is a false dogma that will, in time, have its False Prophet. He will do his very best to make a clear pathway to world-wide godlessness by showing us how, with science and technology, we have no need for a 'saviour'. Unfortunately, Prof. Hawkins' unwittingly played right into that hand when he recently stated that, "Science has made God redundant"; and I have demonstrated that statement is, itself, illogical. The man who can unite the world is only a saviour if the society he brings is better for everyone and that is the key; of course, 'from their fruit you will know them'. Let this book, then, act as a voice crying from the wilderness that makes straight the path to God. Please, teach your children that there is a God and that there is no escaping that God and that science can be used to support this belief. And show them how and why that is the most likely case; if you really get stuck, use my arguments above.

Conclusion

Conclusion

To be fair, this chapter is less of a conclusion and more of a recapitulation of what has been set forth above; yet, it does bring to a close the core component of this book. So, what have we learned? There is more than one kind of truth. A God is omnipresent, omnipotent and omniscient. I proposed that simply by defining one of the Calabi-Yau dimensions as being exactly the Planck Length in length, it defines an eternal link between all the strings that exist and results in the universe being a single object of stringy energy that is omnipresent, omnipotent and omniscient. I then described how I thought energy might be used within the Calabi-Yau space and, in so doing, have made extraordinary claims by hypothesising that 3 dimensions could be used to form consciousness and 3 dimensions used to define the actual definitions of abstract ideas. From this, we could see that the whole unit was perfectly analogous to a CPU, the Central Processing Unit of our computers and that this intelligent 'image' was the basis for the 'likeness and image' we share with God in that, like God but to a lesser extent, we are intelligent and computers are less intelligent than we are. Yet there is, overall, a distinct pattern.

An intelligent entity needs to exist and it needs a type of awareness - we humans prefer to refer to our awareness as 'consciousness', as it sounds superior to simple awareness - and consciousness, as experienced by humans, needs access to abstract ideas. This model of physics is the only model that affords the environment for these distinct aspects of our existence to exist in a way in which seems elegant enough to support our experience of them. Curiously, this model perfectly supports Plato's concept of Forms and the truth of Descartes' assertion that we are, indeed, 'thinking things'. I doubt this to be coincidence and I know, given a space-time continuum, there can be no such thing as a coincidence, because, on a 4 dimensional scale, all events are coincidental. Essentially, Plato and Descartes were on the right track but lacked the physics to back their concepts, yet now we have String Theory, which can support their views.

In form, the Great Entity that is the God of this physics matches with the Kabbalistic Tree of Life and that, too, can be no coincidence. The Jewish Scriptures were always more than just what was written. I formally reveal myself as a believer in the Oral Tradition. The evidence for it is the wealth of information about what was written that could only have come from being discussed at great length over time. The fact that a diagram, the Tree of Life, which outlines exactly what is required for an intelligent being, in my opinion, adds huge amounts of credence to the veracity of a direct revelation from God to the Jewish people. It would have been impossible to give the people at Sinai a message about multi-dimensional space, as proto-Hebrew at that time, simply didn't have the words for it; but a picture speaks a thousand words. The simplicity of the shape itself, purpose driven, lends it the elegance to force us to accept the shape - certainly in light of this model of physics. Quite frankly, it works.

We had a peek into how the fabric of time is a perfect metaphor for how the strings weave together at the quantum level, as so-called 'quantum flux', in my model, is simply energy moving in and out of the Calabi-Yau space. I outlined a view of how a Consciousness Space could exist and why it must exist. I showed why a Platonistic Abstract Data Space must exist and how it might. I suggested that we have a 2-dimensional slice of an overall 3-dimensional consciousness that is God's, which would be a barely imaginable conglomeration of a variety of types of awareness. Yes, God can think like a tree or a blade of grass. And God does so, for God must be omniscient by definition; therefore, God is fully aware and there is no thing that could be more aware than God. This is how agency works and why the responsibility for our actions is ours, as there is only One and we are it.

I offered a reasonable interpretation of how the analogy of God's creative Word can be found to be true when looking at the Periodic Table of Elements, as each period relates directly to a part of speech. This forms a grammar across the table that correlates to molecules representing sentences. A single cell is an enormous book. And then look at the complexity of ourselves. Our quantum story - the quantum story that is each of us - would take billions of millennia to relate, yet it only takes a lifetime to enact. We really should stop to smell the roses.

Then I put forward an argument to solve the question of dark energy. It was a simple case of overlooking what we're looking through. Dark energy is the mass of oblique streams of photons that pass across our focal reference; it is energy we are looking through. We see only those streams that we directly observe, but we have forgotten that all these streams of light must 'pass in the night' across us in ways that are oblique. The more space we look through, the more we find and we find even more in and around galaxies - the stars are streaming photons in ALL directions and we need to account for those that are not aimed directly at our observation point. By putting '$E=MC^2$' and '$E = F\hbar$' together, we can calculate a vast amount of extra energy that would be obvious when we think about it, yet would remain unseen; so our eyes have proven not to be as insightful as our minds can be, because the answer was, literally, refusing to stare us in the face. A fair-minded observer must also account for that which is invisible and that which is both invisible and unobservable.

I presented a mechanism behind the arrow of time. In my model, the fixed, toroidal wall of anti-matter is what draws the elastic space-time through the Higgs structure and, in so doing, moves the past through the present and into the future through the simple power of opposite charge attraction. You have to admit, it's an attractive model. It's simple, all energy is ultimately conserved and a vast array of Big Bang-to-Annihilation cycles is handled with elegance and expertise that is nothing less than Godlike. Knowing the future is fixed is of no matter to us because we are prevented from seeing it. We are left to act in an ever-changing present based purely on our memory of past events. This is truly the harrowing aspect of life and, by far, the most ecstatic.

I then discussed whether or not there was any basis in the claim that science had regarding the impossibility of a six-day Creation. I showed that, when you take into consideration the variability of the Hubble Constant, you cannot, logically or mathematically, rule it out as a possibility. We know that the value of the Hubble Constant during the Inflationary Period was exponentially 99 times greater than that which is required to account for a six-Day Creation. Plus, there is no way of knowing how quickly the 'Constant' slowed. Worse yet for the scientific argument, there is no way of inferring that rate of slowing due to Special Relativity, which, technically, excludes radioactive decay as evidence for the age of the universe; rather in fact, it proves radioactive decay as completely irrelevant because the observed decay would be relative to the value of the Hubble Constant at the time of the observation of the decay and at no other time. So the argument that a six-day Creation is plausible can be based on modern science as we understand it today. Any revelatory evidence could only serve to make it more likely. A God that can reveal how long Creation took and does so is greater than a God who could reveal it but doesn't. And the scriptural revelations of both Judaism and Islam seem to concur that God did, indeed, state that it was six days. Nevertheless, there will always be some scientists bold enough to think they know better than God.

Next we finally stepped firmly into the realms of philosophy - philosophy either derived from the model of physics I present or derived from logic given perfectly reasonable axioms. It is necessary to derive a philosophy in order to know how best to behave, as philosophy must form the basis for a reasonable code

232

of ethics. The requirement for philosophy, in that regard, has never changed nor will it. We must take a view that is reasonable, otherwise we are insane by definition; and we ought to take a view to be as beneficial as possible. But not everyone seems to take that view. So I felt obligated to give a few decent reasons for taking the view that we should be as beneficial as possible. But it will land on deaf ears for those who, by space-time itself if not by God, are destined to harm others.

I decided to offer up a plate of Spinoza for starters - quite literally - as I let his style of argument, a geometrical progression, defend the proposition of 'Creation-ex-Nihilo' - creating something from nothing, which is, without doubt, the ultimate starter. I think I offered a convincing argument even without the tie-in to the Gnostic-side of the Gospel of St. John 1:1. But every little bit of extra evidence adds weight and every connection that we can discover between what we can derive logically and what we are told in Scripture lends credence to the Scripture and to the concept OF Scripture itself. So, naturally, I try to find the links and point them out to the reader.

Next, I tackled a problem that, in some ways, was already addressed - the question of whether or not the universe was teleological - that is, whether the universe was moving towards a known end, an end known by the Creator. Again, I pointed out that Special Relativity answers this question by itself and the answer is, yes, the universe is teleological simply because, in a space-time continuum, all spatio-temporal points or events are already defined within the whole of the continuum. Special Relativity requires us to accept a teleological aspect to the universe.

Then I proffered an argument stating that Newton was not strictly discussing physical bodies in motion with his three Laws of Motion but also, included subtly in his wording of those laws, was that these laws pertained to spiritual bodies in motion as well and used the concept of the gunas of Hinduism to draw the corollary.

As Newton was, first and foremost, an alchemist, he would not have separated the physical and the spiritual as scientists do today but would seek universal laws that can apply to both. I believe he succeeded in that endeavour, yet his success on the spiritual side of the argument has been completely missed by modern science, which baulks at anything spiritual. The blinders of Science, once again, have made us see nothing where, in truth, there was something; therefore, I point out the something that was missed.

Next, I revisited the concept of free will versus fate and, by using only Special Relativity, showed that free will was, for all intents and purposes, an illusion of the highest order. That is, the illusion of free will - the reason we perceive free will - is completely inescapable in that we have the power to perceive potential varying pathways, yet, when we act, we can only ever do one thing at any one time and place and it is those events - in absolute detail, including all the quantum-level spatio-temporal events - that are fixed. They must be fixed in a space-time continuum and, even given the concept of a 'momentum space' as some are espousing currently, the events are still fixed in a continuum of ANY kind by definition. That is the truth and that is the reality. God is the reality and we are nothing more and nothing less than an individual aspect of God's awareness.

I followed that with a demonstrable logic purporting that an omnipotent deity cannot, logically, give away any of its power and remain omnipotent. So the arguments from both science and religion yield the same answer regarding the question of free will versus fate and, once again, science and religion meet on happy terms. Although I suspect that the resulting answer may not be to everyone's liking that is not a factor when discussing truth. Truth is not a thing that is only held when we approve of it; rather, we must embrace the truth irrespective of whether or not we like it. Could you be content knowing that you believe a lie? I couldn't. Yet the actual fact of the illusion of free will is such that we are still left with having the choices we perceive and, effectively, the illusion is

234

just as good as the real thing. Nothing short of an omnipotent and omniscient God could set up such a perfect illusion.

Then came the question from left field, how does emotion fit into this system? The answer was that it is this very mechanism that affords us the chance to act beneficially. Our feelings may boil up inside of us uncontrollably but we have some time to decide how we demonstrate (or emote) those feelings. This is our golden opportunity to act wisely and not only affect the future for the better but, perchance, also alter the past through forgiveness. This is real forgiveness, though, and nothing short of it; true forgiveness removes completely any ill feelings regarding an action. It alters how we perceive a past event. So, although we know what happened and how it hurt, we let the hurt go because we understand that, in this universe, events occur and some we will adore and some we will abhor. In either and any case, we 'must needs' accept that which has happened and move forward with everyone's best interest in mind in order to be the best us we can be. Acting wisely, which can only be done when we know the truth, is the closest thing that humans have to being powerful in any sense because acting wisely is acting Godly.

I then spoke about how this philosophy, especially with regard to the new found view on free will, affects our criminal justice system. I explained how it changed from holding a person responsible for 'murder' to holding them responsible for not having perceived that they did not have to act in such a way as to cause the death. We accept that the event of murder has occurred but we cannot condone the act itself; therefore we react by punishing the individual perceived not as having caused the act - because we know that God causes all actions - but as the individual who did not think of a non-criminal way to behave. Thus, our justice systems can remain pretty much the same; what changes is that the public and the system have a greater understanding of the nature of events in a space-time continuum and both the public and the system know that, in the inevitable event of our failure to impart true justice, God will invariably enact perfect justice inevitably and we should put our trust in God's justice if and when we see our pathetic human attempt at justice fail. God's justice is as fair as 'equal and opposite reaction' and is just as real. We must also remember that God's mercy always prevails over his wrath but this does not affect perfect justice in any way.

After changing the basis of our understanding of justice I rent the underworld asunder and stated plainly what sin was and how damnation works. Straight into the pits of Hell we must go to see if our philosophy can withstand the inspection. Sin is vanity and selfishness. These are the traits that will be punished because they act, in their own way, as a denial of the oneness of God and that is the greatest blasphemy in a monistic universe: to deny the oneness of the whole. To act in such a way as to be seen or perceived to be vain or selfish is the means by which we earn disdain before God's metaphorical eyes. When the truth is that there is only one thing that exists, anything that makes an individual stand out as acting only for their own interests will mark them as uncivilised in a civilisation that understands their place in this system. We are all in this together and we ought to work together. When we do not, then we separate ourselves from one another and the end result is a feeling of separation from God's presence in the afterlife. In essence, we damn ourselves but not without earning it and having it all explained to us by God. And we learned, in most likelihood, that only a single damnation will be eternal and that is reserved for the entity called Satan or Iblis; some humans may be damned for a long 'space-time time' but will, most likely, not be damned eternally, for no lesson could be learned and that is not in keeping with the formula that God's mercy will prevail over Its wrath. Nevertheless, all 'time' spent not attached to space-time will seem eternal for those who simply exist in the timeless afterlife. On the good side, it is also logically derived that rewards in Heaven will be truly eternal as anything less would not be in keeping with God's attribute of being 'Most Merciful'.

I then discussed the potential process for the resurrection of the dead. I proposed two possible scenarios: one which is more like a word-processor style 'copy-n-paste' method and the other more like a kind of *Star Trek* transportation, where some molecules get reassembled into an entity that looks and is identical to the original. Either way, the link between body and soul is re-connected for a brief time before the annihilation sequence that ends every Big-Bang-to-Annihilation cycle. Because we do not know when this annihilation will take place, it is in everyone's best interests to act as if it were tomorrow - if not later today. In other words, we should try to reconcile ourselves with anyone we are not reconciled to.

This naturally led into the final two chapters on how we should act and why. The end philosophy is nothing new. Do unto others as you would have others do unto you. Love your neighbour as yourself. Love God with all your heart, all your soul and all your might. These timeless phrases are the philosophies of the greatest of human prophets and, if this model of physics is a true representation of reality, then now we know why those phrases have been given to us to guide us. There is perfect logic behind them and that is why an omniscient and omnipotent God would send prophets to mankind with those very messages. Think how blessed we are to have had just a few revelations from God. It is because God is Most Merciful that it is incumbent upon that great entity to give us the opportunity to mend our ways and to act less selfishly and vainly and to act more selflessly and Godly. Ultimately, we ought to try to be the best that we can be and that means being as empathetic and as sympathetic to others as is humanly possible. Anything short of that and we have failed in our attempt. But, should we stumble, we know that God is Most Merciful and that if we meant well, then that is what will be taken into account. The main things for which we will be held

237

accountable are our lifelong attempt to be selfless - to help our fellow life forms as best we can - and to make as small as possible the detrimental knock-on affects of our living on our living and non-living environment.

This is a huge responsibility and is the core of the philosophy that can be derived from this system of physics. We ought to act together for everyone's best interests and always try to act for the greater good. The hardest aspect of that, though, is that we may not be able to perceive which act, at any given time, is the best act. Well, we can only do our best and we know that God will understand our intentions. So, we ought to have that perfect faith in God too.

Science is not opposed to religion; it never was and it never will be. Science is the attempt to discover how God created the universe and, if we can, peek into the possibilities of why God created this universe. Only through spiritual aspiration and recognition of the spiritual aspirations of one another can we achieve the step forward that is possible by accepting this model of physics. This model tends towards confirmation that there is, indeed, a God and that that God has a plan for the universe and that we humans are a vital part of that plan. In this model, science meets religion on happy terms and both are satisfied with their relationship. No other model provides the answers to so many questions and no other model comes close to explaining any of the 'whys' we humans have regarding the nature of existence. I only hope that I have relayed this as clearly as I can.

The next thing is to try to apply this philosophy in some real-world situations and see if we can make a difference. The 'Practical and Social Applications' section of this book is very political. I point out inconsistencies within certain religions and I offer advice to certain countries. I put forward a potential policy for peace in the Middle East (Israel/Palestine). I lay down the ways and means towards establishing a Caliphate for the Islamic nations of the world. I also state harsh criticisms to some countries that are desperately in need of help. Finally, I finish off with a couple of exegeses in which I raise certain outside-the-box ways of thinking about certain topics.

So, the main part of this work is at an end and it is time to apply the philosophies. The rest of this book speaks to many specific audiences; however, each reader will fall into one of the categories addressed, so, do read on, as there is more for you. Plus, you will see how, when these philosophies are applied, the changes that could be made in this world are truly amazing. Please, dear reader, be a part of the change for the better that we could all embrace if we take on board the philosophies presented here. Each of us is vital and we must acknowledge our fellow humans or we will suffer at our own hands for not doing so when we could have. It is SO incredibly simple; yet SO complex. That is the nature of truth in a universe where there is truly only One. Accept the oneness of our universe and be the best you can be. Do it for God; do it for your loved ones; do it for the sake of those you do not know and do it for those who you may feel are your enemies. Lastly, do it for yourself but only in the knowledge that there is no difference between you and the next human for only then is it not really selfish because you will understand that we are all extensions of a greater One that is none other than God.

Practical and Social Applications

Advice to Religions

Judaism

Historically, it has been extremely difficult to tell the Jewish people what to do unless, of course, you are God and, even then, they still backslide. If you are not God and you tell them what to do, they will tell you where to go. I give advice that is based on an understanding of the truth of the story of Judaism. I greatly respect the faith; but there have, as you know and as has been recorded, several lapses in the practice of the faith for various reasons. There are lapses today and they should be corrected from within. I hope that my advice will be taken, as I believe it will lead Judaism to the pinnacle of its existence. These few people - the Jews - form an incredibly small proportion of the world's population; but, the amount of difference these people have made to humanity historically is more than any other people of similar population. This is no coincidence, as God has always had a purpose for these Sons of Abraham. But you, the Jewish people, must do as God has instructed you to do or you risk losing everything... again!

There are several things which you must do. Firstly - and this will meet with harsh criticism by many faithful - is to understand that it is most likely that the Torah, as it stands today, is NOT the same document that was revealed to Moses and the Israelite tribes at Sinai. The scientific and scholarly evidence points to the likelihood that the Torah, as it exists today, was compiled from four source documents: The Yahwist Scroll, The Elohist Scroll, The Priestly Scroll and the Deuteronomist's work. It is also likely that the Prophet Ezra was the individual who compiled these into the current Torah and that his hand is, in fact, the Deuteronomist's hand. Although, to his credit, Ezra did a beautiful job and the work he produced was truly inspired and contains all the essential elements of the original; plus, I believe firmly that Ezra's heart was in the right place and that he compiled the Torah for all the right reasons. For example, the people wanted to hear it, the religion required it and, by the law itself, it needed to exist. Many will be shocked by this, but the evidence is hard to argue against.

For the record, here's what I think happened. Firstly, the Torah was revealed to Moses and, most probably, to the entire generation of Israelite tribes at Sinai. This revelation was written on scrolls and placed in the Ark of the Covenant with the second Tablets of Law, the pot of Manna and Aaron's rod. When the Philistines (Plishtim) stole the Ark from Shiloh and took it back to their homeland, they opened it and it unleashed a firestorm that not only killed scores of Philistines but, almost assuredly, anything flammable within the Ark and THAT would have included the scrolls of the Torah, the Manna and Aaron's rod. The Philistines returned the Ark and, later it was moved into the Temple of Solomon. Solomon had the Ark opened and, once again, it burst forth, but only killed a couple of people nearby. When Solomon looked into the Ark, there were only the Tablets of the Law left in it. This is all recorded in Jewish scriptures. When Solomon saw that the Torah was gone, he set his scribes and scholars to work on recovering it via the memories of the most learned people to which he had access - the Scribes and Priests - and they wrote the Yahwist Scroll, the Elohist Scroll and the Priestly Scroll. When the Babylonians invaded the Kingdom of Judah, they took all these back to Babylon not knowing or caring what they were. Yet the Scribes, Priests and rabbis were aware and kept them as safe as they could. When Cyrus of Persia invaded and defeated the Babylonians, he allowed the Jews to return to Jerusalem and the former Kingdom of Judah and helped pay for the rebuilding of the Temple, as is well known.

This was in the time of the Prophet Ezra and he was one of those learned scholars who knew that the only source scrolls were NOT the originals but, at least, they were the best they had. He knew that the people would demand the Torah and he knew that the religion required it. So, bravely, he took it upon himself, as a recognised prophet of the faith, to put these scrolls together into a cohesive whole and declared it 'The Torah'. But, when he read it to the people, they balked at the mentioning of having to live in a Succah (booth) at Succoth and remarked that they had never heard that before. This, too, is recorded in Jewish scriptures. So there is evidence that the peoples' memories of the Torah differed from the work Ezra produced. And, unfortunately, this was the beginning of the questioning of the Torah's 'correctness'. As I said, I believe wholeheartedly that Ezra acted in the best interests of his people and his faith; however, he could NOT reproduce, exactly, what had been in the original revelation at Sinai, as so much had happened between his time and Moses'.

In my opinion - and it is only just that - the original Torah told the story of Creation, the Antediluvian patriarchs, Noah and the flood, the stories of Abraham and his sons and grandsons all the way up to the bondage in Egypt. In other words, the Torah contained most, if not all, of Genesis. It also would have contained the first 20 chapters of Exodus, as this tells the tale from the bondage to the revelation at Sinai. I also believe that it must have included the laws of kashrut; the design for the Ark of the Covenant and the Sanctuary around it; the laws pertaining to the priests and sacrifices, cleanliness and certain benedictions; it also included such concepts as Tzitzit, Tefillin and Mezuzah; the laws concerning the Sabbath including what constituted work; the laws pertaining to Yom Kippur; the forbidding of usury of any kind; the importance of circumcision for males and, without doubt, the core teaching of the Shema ("Hear, O Israel, the Lord Our God, the Lord is One").

As the laws of kashrut are outlined in Deuteronomy, this could imply that Ezra realised that the previous Scribes and Priests had forgotten to include them in their scrolls. I refuse to believe they were not revealed at Sinai based on the evidence that dietary laws for Muslims were included in the Qur'an - I can see no reason why the more stringent laws of kashrut would not have been included in the Torah, which is far stricter, generally, than the Qur'an. That said, this would not include any of the rabbinical interpretations of these laws, as most of the rabbinical interpretations regarding kashrut are designed as a part of the structure known as 'the Fence around the Torah' and, strictly speaking from a literal or 'pshat' interpretation of the Torah, are not actually binding; however, the 'Fence' does prevent Jews from transgressing the laws of kashrut and that was the reason for those interpretations, as they were based on the good intentions of the rabbis.

I do not believe that the Torah would have included anything regarding the 9th of Av or Hanukkah, which it does not even in today's Torah. Those celebrations were borne out of respect for later events and their every aspect is purely rabbinical in origin; this is, indeed, well known. But far more importantly, it would not have contained any historical events AFTER Sinai. So, those historical events listed in the latter part of Exodus, Leviticus, Numbers and Deuteronomy - especially the death of Moses himself - would NOT have been in the original Torah.

So, what does this mean, then, for Jews? It means that Hanukkah, for example, is not obligatory, at least it has no authority via the Torah; rather, its authority lies with latter rabbis. The same holds true for fasting on the 9th of Av. But it does demand that the Jews have an Ark of the Covenant and a Sanctuary for it, as they were commanded to have an Ark of the Covenant and a Sanctuary for it. If an Ark is lost or stolen, they should recreate one, because they were commanded to HAVE one. Likewise, if a Sanctuary is destroyed, they should recreate one.

This is a very important facet, as it plays a large role in Jewish eschatology. Allow me to explain.

Without the Ark, there is no requirement for a Sanctuary or Temple, as the Sanctuary was only created to give a sacred place AROUND the Ark. The Ark is more important than the Sanctuary. The Messiah is supposed to rebuild the Temple, which serves as a more permanent Sanctuary, but there will be no reason to do that if there is no Ark! Therefore, I implore the Jewish leaders to recreate the Ark according to the design that was put forth as well as recreating a Sanctuary for it; for then, and ONLY then, will the Messiah have a reason to rebuild a more permanent Sanctuary, a Temple, around it. Plus, there is no Torah-based requirement for the Sanctuary to be on the Temple Mount, as history has shown that Shiloh, for example, was perfectly acceptable to God.

Remember the words regarding Judah in Genesis 49:10, "The staff shall not depart from Judah, nor the sceptre from between his feet, until Shiloh come, and the obedience of the people is unto him." I believe this to be a true line from the original Torah and it implies that the Jewish people need to have an Ark of the Covenant ready - at Shiloh - as a proof of their faith that their Messiah will come. And I believe that both Christians and Muslims alike should endeavour to help the Jewish people make this a reality, for neither of these faiths have anything to fear from the Ark of the God in which they all purport to accept and believe.

I do believe, though, that by not having an Ark, Jews prevent the coming of their Messiah and that through their own idleness in this regard have reneged on the Covenant they so willingly embraced at Sinai, which bade them to have one. Whilst it is true, the Torah never mentions what to do if an Ark is lost or stolen, the fact remains the Jews were commanded to have one; so, I believe that implies either finding the lost one or, if that fails - and I think by this time we can say that it has - to rebuild one in order to maintain the existence of one. If you want to prove me wrong in this regard, then prove me wrong by building it and see whether or not the people are rewarded for their faith!

I don't know if the Israeli government, as it is a secular government, should be involved in the re-creation of the Ark, if they wish to have 'overseers' present, this is fine, but the Orthodox Community should realise the importance and immanence of what I say. I cannot stress more the importance of having an Ark. Jews were commanded to have one and have not had one for nearly 2000 years. This is the greatest lapse of faith that Judaism has ever seen in its entire history and it must be corrected as soon as possible. I believe it will give far greater hope to the Jewish Community worldwide than saying, "Next year, in Jerusalem" ever has.

If this aids the coming of the Messiah you so desire, you must realise that, by anointing any man as the Messiah, your government is set on a course towards a monarchy rather than the parliamentary democracy it currently enjoys. Can the people endure a monarchy? Will they accept it? If they wish to EVER accept a Messiah, then they MUST accept, for his lifetime, a monarchy. And, of course, who will anoint this man? It is to be done by a recognised prophet or, at least a 'Holy Man' recognised by all Jews as authoritative. The Qur'an, which you reject, says there will be no more prophets. This causes a small problem and leaves you with little choice but to declare an individual as a 'Sacred Authority'. This will avoid problems with the Muslim Community by whom you are surrounded, as they will not tolerate the declaration of a prophet but will have no problem with the declaration of a 'Sacred Authority' for the Jewish people and it is a peaceful way forward, which is what the Torah commands, if possible, and what God expects.

Because the anointing of an individual as 'Messiah' implies a return to a monarchical form of government, I propose that a law be passed by the Israeli people that will allow for this. This monarchical government need not be a permanent form of government but should only last as long as the 'Messiah' lives. Thus the law should maintain that the government be returned to its current form upon the death of the 'Messiah'. However, taking Jewish eschatology into account, it is entirely likely that Armageddon, the Resurrection of the Dead and the Final Judgement of this world will happen before the 'Messiah' dies. So, there may be, in fact, no turning back once an individual has been anointed as 'Messiah'. This fact, though, should not deter the current government from passing a law that allows for the coming of the Messiah. It ought to stand as a demonstrable proof that the current secular government embraces the Jewish faith and will not stand in the way of the requirements of the Jewish faith. If it refuses to pass such a law, then that could only be interpreted as anti-Semitic and show the government of Israel to be hypocritical and uncaring with regards to their populace and their beliefs.

I recommend that the leader of the Orthodox Community in Jerusalem, or whoever would be the one who directs the re-creation of the Ark, be declared this 'Sacred Authority' - whoever it may be at the time. I say this as I feel that it is the Orthodox sect that would be the most likely to ensure that all the laws concerning the creation of the Ark of the Covenant would be followed. I would hope that the Chief Rabbi of Jerusalem would take it upon himself to bring the faithful back to the faith and lead them by demonstrating his faith by instructing the re-creation of the Ark; but, of course, no man or woman should be forced into taking this responsibility. Rather, it must be felt by the individual to be a self-imposed obligation and there is no reason why this individual must be a man. Devorah would have done so in her time as would Miriam in the absence of her brothers Aaron and Moses and, I believe, if someone had informed Golda Meir of this situation, she would not have hesitated with regards to the endeavour. God has always been egalitarian when it comes to leading Israel forward and, while there have been fewer women than men who have done this, the women who have served as leaders have always been incredibly remarkable and memorable... always.

Israel is surrounded - surrounded by Muslims on all sides. There are some who believe that this means that they are, in fact, surrounded by enemies; but, this is a lie. There are those - on both sides - who do not want peace; they are neither truly Jewish nor truly Muslim, as good Jews and good Muslims always strive towards peace. Whilst it may be true to say there is enmity, the people themselves are not enemies. They are brother nations. The Palestinians of today are NOT the same people as the Philistines of ancient times. These modern-day Palestinians are Arabs; they are not the genetic descendants of the Canaanite Plishtim. Arabs and Jews used to get along fine but, then, the pot got stirred. The UN's division of Palestine was the beginning of the bad times that came after. The real Jews and Muslims - those who want peace - must work together and do so openly. Those who do not want peace will then show themselves and they can be rooted out. Those who do not want peace, be they Muslim or Jew or other, should be discovered and dealt with justly by those Muslims and Jews who are working together towards peace.

Below, in the 'Advice to Islam' section, I have suggested a potential solution to what has been termed 'The Palestinian Problem' that, I believe, would end the enmity for all time. I implore the Jewish readers of this book to read that section too and work towards its goals. It is no longer a matter of 'who started it'; what matters is that the enmity is ended by good and honest, peace-loving Jews and Muslims. To have peace restored to the Middle East will, naturally, bring less tension worldwide. The entire world should help where it can and Israel should welcome any help - especially if it comes from the Palestinians themselves. They, like you, have a right to a homeland and sovereignty and the two must come to terms. I believe that my idea, below, will reap such a reward. Armageddon may well be the 'war to end all wars' but it needn't be a hugely destructive war; rather, it could simply be the combined effort to fight those who do not want peace in the Middle East. Please, keep this interpretation of Armageddon in mind.

There are many so-called terrorists on both sides that make the Palestinian Problem a reason for their actions and use aggression as an excuse to further their own war-bent purposes. So, pull the carpet out from under them and stop all aggression towards the Palestinians and vice versa. Be brave. Work together for peace. Moses' own wife was a Midianite, an Arab; so, you must know that, in the past, there was no enmity between these peoples and there is no good reason for it now. They are - you are - brother nations from Isaac and Ishmael, the sons of Abraham. You have the power within you to become, once again, a united family, even if you have differing faiths. God approves of both Judaism and Islam, as God revealed them both. If God approves of them both, why can't the practitioners of those faiths tolerate one another? History proves that they used to and, thus, they can again. Be brave and follow your faith; this may be hard but it is worth doing and can never lead you astray.

Christianity

Christians, your greatest threat, today, comes from within: the right-wing, Neo-Con fundamentalists. They told you that Osama Bin Laden was a typical example of a Muslim. This is as true as stating that Hitler was a perfect example of a Christian. As we all know, nothing could be further from the truth. These fundamentalists want to bring about the end-times. They want Armageddon. Their goal is to bring about a war between Christians and Muslims, a Final Crusade that they think Christians would win. Well, if the original series of Crusades, of which the Christians only ever won one - all the rest were won by the Muslims - are anything to go by, then they will lose. These fundamentalists want to stir up trouble between people to bring about this war that they believe will be Armageddon. Why, in God's name, would anyone in their right mind want to bring about what, in essence, would be World War III? This work is done, mostly, in America, where the principles of 'Freedom of Religion' and 'Freedom of Speech' give them a platform from which to spin their yarns; but they abuse these principles as neither 'Freedom of Religion' nor 'Freedom of Speech' were ever intended to act as a framework to support the spreading of lies and hate. These fundamentalists are pulling the strings behind the Conservative politics and are, slowly but surely, infiltrating the House and Senate. The White House is their goal. From the Oval Office, one simple executive command can start the REAL 'war to end all wars'; but, having it their way could end in the slaughter of millions of innocent people and all for the bloodlust of a few deranged 'believers' that are, in essence, no different than the recent deranged Christians like Jim Jones or David Koresh. They don't realise that the White House has its own guard and has had it for years. This is a mighty struggle between dynasties. Heck, the White House is just a building; what matters is 'who is calling the shots'. In my 'Advice to The United States', below, I tell you who's calling the shots.

Christianity as a Global Community must not allow them to succeed. They are a fascist hoard within Christianity that must be weeded out in order for the best of the faith to survive. They are just like their Islamic counterparts, the various Islamist terrorist factions; these people are not good Christians just as anyone who straps a bomb to themselves and kills innocent people is not a good Muslim. Beware their camouflaged fronts like 'The Tea Party' that tries to remind you that these people are, like Samuel Adams, real believers of freedom and all that for which America stands. In truth, Samuel Adams was a black-marketeer and threw the British tea into the harbour because the British had LOWERED their taxes and undercut Adams' black market profits. THAT was the reason for the 'Boston Tea Party' - to rid Adams of his competition and leave the colonists with no choice but to buy his black market tea. How very monopolist of him; and certainly not in a spirit of fair play neither much regard but for greed. Today's 'Tea Party' is just as much of a ruse. It wants to take the Constitution and true American ideals and principles and throw them into the harbour. They say that America is a 'Christian Nation'. This is a lie. America was founded to support no specific religion; it was founded to offer a safe harbour for all faiths because the founders knew that it was wrong to persecute an individual simply for having different religious beliefs. The Tea Party's platform is quite insane, really.

I realise this has become a bit of a tirade; but America is threatened by the most wicked and insidious of Christian varieties since the Inquisition and the people of America, as well as the people of the world, need to be aware of it if they aren't already. Any 'group' who speaks of 'pure American' anything should set off alarm bells. The 'pure Americans' - if there are or were any - are the Native Americans, who were ruthlessly dealt with by the American government as history proves - to the point of biological warfare in order to cause genocide. They were 'given' land that America didn't want, sometimes forced to walk hundreds of miles to it and then 'allowed' to govern themselves. These Indian Reservations are some of the safest places in America these days. The Native Americans know what it is like to be treated badly simply for living differently and believing differently; please, learn from that historical mistake and do not let another government repeat those ghastly acts on any Americans or the people of any other nation.

I think I've spoken enough on that, now, and it's time to return to religion generally and to speak to Christians globally rather than focusing on American Christians. This, like many of my statements to other faiths, will come as a hard teaching and will make many people angry with me because I have provoked them with thoughts; but, please, hear me out, as I only have your best interests at heart. You must re-address the concept of what is Gospel. In order to have a firm foundation that is truly Christian - that is, based on the teachings of Jesus of Nazareth - the faith must be over-hauled and renewed. This is not easy to do and it means ridding the Christian Scriptures of most of Paul's teachings.

Paul, Saul of Tarsus, was a man who never met Jesus. Additionally, he avoided meeting any of the apostles for three years (gladly admitted in Galatians 1:17-18) while taking that time to spread his own version of the events and their meaning and, when he could avoid them no longer, he only met with Peter and James, the brother of Jesus; he still avoided the others. The Apostles were, rightly so, angered by his actions and his teachings, as he had taught many things that Jesus never taught. Jesus never said that he was part of a trinity. The trinity concept was used by Paul to get Greeks to accept the faith, as it was a concept to which they were accustomed. Similarly, the entire Pauline concept of the Eucharist was taken from Mithraism. A worshipper of Mithra, like most of the Roman soldiers of Tarsus were where Saul grew up, would take a bull dedicated to Mithra and eat its flesh and drink its blood in order to become one with Mithra. If the worshipper was too poor to afford a bull, they were allowed to use bread and wine instead;

the bread would substitute for the flesh and the wine would substitute for the blood.

When Paul heard the story of the 'Last Supper' by early Christians in Damascus, he invented a new way to bring in more believers - by making a Mithraic ritual appear Christian. Remember that, in all likelihood, Jesus had only said, "Remember me when you eat and drink". Anything more like the story of the blood of the 'New Covenant' and Jesus' insistence on the apostles' eating his flesh and drinking his blood were based on Paul's concepts and not on Jesus' original intentions. Jews are specifically commanded NOT to partake of blood in any way; this is a vital part of the laws of kashrut, the laws that make food kosher, i.e. allowable for Jews. I find it incredibly difficult to believe that Jesus, as a rabbi, would ever teach people to even imagine they were drinking blood.

These changes that Paul instituted were far-reaching. He decided that circumcision was no longer required when Jesus had never said anything of the kind. Paul states clearly that it is HIS Gospel in which Jesus is the Son of God. Peter, Andrew and James were incredibly incensed by the audacity with which Paul changed things to suit himself and they were right to challenge him; but, unfortunately, the damage had already been done. Eventually, the Apostles relented as they realised two things: Paul had spread his version so far that it could never be recovered and it resulted in many people embracing the faith. Because of these two facts, the Apostles realised that Paul's mission was resulting in Christianity growing but at what cost? Peter never trusted Paul nor did James. They avoided contact and each tried their best to instruct believers in their own way.

But it must not be forgotten that it was Peter who was supposed to be the foundation for the continuity of the faith after Jesus' death. Jesus made this clear when he said, in front of all the Apostles, in Matthew 16:17-18, "Makarios ei, Simon Bar Jonas... su ei Petros, kai epi tautay te Petra, oikodomayso mou tayn ekklaysian", which translated from Greek means: "Blessed are you

255

Simon, son of Jonah... you are Peter and on this rock I will build my preaching". Now, this word 'ekklaysian' is usually translated as 'Church' but this is an incorrect translation. There is no word for 'Church' in Koine Greek (the dialect of Greek in which the Gospels were written), as the concept of a Christian Church didn't exist at the time; so, the translation of this word as 'Church' is completely anachronistic, misleading and wrong. 'Ekklaysian' means 'preaching' just as the book of Ecclesiastes is sometimes called 'The Preacher'. It was Jesus' intention that Peter be the one who carried on the true teachings because of Peter's knowledge and faith. Yet it was Paul who usurped this post and took it upon himself to spread his own version of 'the truth'.

Jesus' mission was to encourage Jews to live by the spirit of the Torah rather than by its literal word. This was because the Sadducees were in control of the Sanhedrin in Jesus' time and the Sadducees only accepted the written Torah and the literal meaning of its words. Whilst he agreed there must be a new understanding of the Torah - his 'preaching' - he did not encourage people to reject the Torah or its Laws. In fact, he stated plainly that nothing 'not one jot or tittle' from the Law (Torah) would change. Therefore, it is clear that he would never have permitted the renunciation of circumcision by Paul, as it went against his teaching. Jesus probably realised that, in time, his teaching may form the basis of a new 'sect' of Judaism that would have a far better understanding of the Torah than either the Sadducees, who only accepted the literal interpretation, or the Pharisees, who didn't practice what they preached and, therefore, set a bad example. I do not believe, though, that he intended the followers of his teachings to be anything other than Jewish. Paul, though, saw things differently. And we ought to remember that Paul was from the tribe of Benjamin and was, therefore, likely to suffer the old Benjaminite prejudice against Judah that arose after David, of the tribe of Judah, succeeded Saul, of the tribe of Benjamin, to the throne of Israel.

Jesus' vision would probably have been very similar to the Conservative/Masorti style of Jewish worship seen today, which is based on an understanding of how ancient principles can and must be employed to serve a modern world. Although I suspect Jesus' interpretation may have had a slightly more 'Reform' attitude - only to keep the faith fresh and strong - but nowhere near as lax in orthodox practice as modern Reform or Liberal Judaism. Paul envisioned a completely new faith carved out of the old with him, Paul, as the main founder... in Jesus' name, of course, as Paul had no real authority. It was Paul's vision that came true and Jesus' desire that Peter's leadership be the true authority was totally eradicated by Paul.

Because of the above facts, I would suggest that a new council be arranged by the Pope and Cardinals and that a new 'New Testament' be authorised and that it excludes most if not all of Paul's writings, puts greater emphasis on the writings of the other Apostles and includes some of the other Apostle's 'Gospels', even if there is current debate regarding whether or not any Gospel was actually written by that Apostle whose name it bears. The test is whether or not the Gospel in question maintains Jesus' teachings or not. The point is that the 'teaching' must match what Jesus himself taught and that the resultant Scripture would be far more consistent with itself than the current New Testament is. Self-consistency is a major attribute of the truth and neither God nor his true prophets are liars. The truth is always consistent with itself. For example, why would the 'Prince of Peace' "come with a sword"? This makes little sense, unless you accept Paul's teaching over Jesus'.

A true Christian follows the teaching of Jesus not the teaching of Paul - that would be a Paulist - and there are far too many Paulist tendencies (one is too many!) in Christianity than there ought to be. There should be no such concept as a 'Christian Soldier marching off to war' when Jesus instructed people to 'turn the other cheek' and 'Love your enemy'. Consistency is the key;

even the supposedly 'synoptic' (meaning 'looks the same') Gospels of Matthew and Luke don't agree on the genealogy of Joseph back to David! Surely that is proof that something is wrong. It is more likely that Matthew's is correct, although his knowledge would have been second-hand at best; Luke's knowledge would have come from Paul whose knowledge was either self-concocted or would have been second-hand, thus making Luke's knowledge third-hand at best.

I know that which I've said above will anger many and come as a shock to many; but listen to what I'm saying: I'm saying, "Follow the words of Jesus". This is not an anti-Christian message and I am not, in the least, anti-Christian; but, I would have more long arguments with Paul than I would with almost any man from history. I would stand next to Peter and Andrew and James and argue with them against Paul. I view Paul as "the used-car salesman of Christ"; he would tell anyone anything just to get them to buy his story. As far as I'm concerned, he stated clearly that it was all right to lie for the benefit of God's greater glory, which is, of course, a blatant absurdity given the commandment to not bear false witness. Paul stated in Romans 3:5-7, "Now if our unrighteousness commends God's righteousness, what shall we say? That God inflicts wrath on the unrighteous? That is 'according to a man' I say. May it not be (like that). Otherwise, how will God judge the world? But, if God's truth is abounded by my lie, to His further glory, why, still, am I also judged as a sinner?" In other words, Paul felt that, if a person lied in order to give greater glory to God, then the lie was justifiable. This teaching is a lie because no lie is justifiable. "Thou shalt not bear false witness." It's as simple as that; yet Paul had to justify his teachings because they varied from the teachings of Jesus and the Apostles and, obviously, they varied with common sense and with truth itself. Truly, I believe Paul was a very troubled man.

Let us not overlook the fact that Paul was also a misogynist. According to the Gospels of Thomas and the Gospel of

Mary (Magdalene), Jesus confided and entrusted many of his most secret teachings to Mary Magdalene; whereas, Paul says, in 1st Timothy 2:12, "....I do not permit a woman to teach, nor to exercise authority over a man..." Who, in God's Name, did he think he was? I believe Jesus would have been appalled by this attitude. God created both men and women and both are equally required to continue the species. Neither is superior to the other except, perhaps, by piety alone. The effectiveness of women as school teachers is perfectly accepted in today's world; yet, Paul would not allow even this. And a son who did not yield authority to his mother and learn from her would obviously be dishonouring her and breaking yet another of the Ten Commandments. Yet, Paul would not permit any woman to teach any male. Why? According to Paul, it's because of Eve and the 'fact' that Eve succumbed to temptation before Adam. The truth is they BOTH succumbed and the timing is irrelevant. Paul didn't have a wife and I can understand why; if he did, she died long before he watched Stephen stoned to death or he abandoned her in order to preach, as there is no mention of her in any accepted Scripture. With that attitude, he couldn't afford to have a wife because, if they had male children, he would have to keep them away from their mother to protect them from her because she might accidentally try to instruct them with regards to anything. Therefore, I make the presumption that he had no wife and/or no woman could bear him as a husband.

Personally, I think Paul was very guilt-ridden regarding his persecution of early Christians and was guided more by his guilt than his supposed 'new found faith'. He had to convince the Christians and others that, really, he wasn't a cold-hearted killer of people who thought differently - even though his early history stands against him in that regard. I believe he truly DID feel guilty about that and set out to try to right his own wrong. Unfortunately, he compounded his blood-guilt with lies and false teachings and is responsible for leading so many millions astray -

those that believe him - that I dare not think what has become of him in the afterlife. I also believe he was a bit of a hypochondriac. Why else would he have his most faithful companion, Luke, be a physician? Surely, Paul trusted God enough that his faith could have withstood hundreds of fifteen-foot high piles of bacteria; but, no, he had to have a personal physician with him as often as possible. That doesn't sound like a man with great amounts of faith; rather, it sounds like a man in desperate need of someone in whom he can confide and someone who can try to heal him physically and mentally/psychologically.

Please, I implore you, Christians, to rethink the value of Paul's teachings and to disregard them when they differ from those of Jesus. Only then can you truthfully call yourself Christian. Remember that Jesus was born and lived as a Jew. There is a very strong argument, though, that, technically, he died a Muslim. By Muslim, I do not mean a follower of the Prophet Mohammed (saw) but one who completely submitted his will to God. In Gethsemane, Jesus clearly stated, "Not my will but Thine be done." This is evidence that Jesus clearly submitted to God's will and, de facto, became a Muslim and, therefore, died as a Muslim (if you accept that he died on the cross. If not, then he ascended to Heaven as a Muslim). He accepted that God's will is the only will that prevails and that understanding should be handed down as one of his last and greatest teachings - even though it is clearly a Muslim teaching. Jesus understood that an omnipotent being has complete control and we are only God's agents performing God's will. This is a monistic and mystical understanding, which is evidence that his teachings were based on a monistic understanding of God. And, now that we have Special Relativity, that belief is, now, a scientific fact given the existence of a space-time continuum.

Also, understand that Muslims are not an enemy to Christians nor have they ever been. Those who would lead you to believe otherwise are liars and, in fact, are anti-Christian themselves. According to Islam, Christians are a protected people (Dhimmu), as are Jews and Muslims are forbidden to provoke them and forbidden to fight against them unless it is in self-defence. Likewise, it is un-Christian to provoke anyone; rather, you are instructed to treat others as you would be treated and to love your neighbour as yourself. It's high time that Christians took this seriously, as, once they do, the world will be a much safer and happier place in which to live.

So, I beseech Christians as a whole and I beseech the Pope in particular, please have a serious re-think regarding Paul's teachings and how they have infected the tenets of true Christianity by how much they differ from the teachings of Jesus. If you do so, Christianity will have far greater internal consistency and, therefore, it would be a stronger faith for it. If you do not, then you may as well call yourselves Paulists and stop lying to yourselves and the rest of the world.

Islam

There are both political and theological considerations to be made in order to improve. And, of course, the continuing 'Palestinian Problem', which, by its very nature, is a combination of political and religious aspects and must be treated in isolation from the general political and religious advice. Therefore, I will treat these areas separately, although I recognise there is an implicit overlap between them.

... Political

Firstly, and this is going to be very hard to do, you must drop the borders between Islamic nations. There was never a time in the past when hard and fast borders have ever been useful to Islam. Most previous Empires - like the Ummayad, Abbasid, Ottoman, Safavid and Mughal - did not have strict borders like the modern Western concept of a Nation State. How can there be borders WITHIN Dar-al-Islam? It is anathema to Islam as it was practiced in the time of the Prophet (saw) to have such strict borders; therefore, to have such borders is not following the Sunnah of the Prophet (saw). Furthermore, it constricts the free movement of Muslims to wander and live throughout Dar-al-Islam. Simply put, nation states are not compatible with Islam.

There are, of course, countries like Bangladesh that are completely surrounded by non-Muslim nations and these must be handled as Emirates. In fact, in order to compromise with today's structures, all modern Islamic nations could retain a level of regional sovereignty by regarding their current leaders, at the time of transition, as Emirs governing an Emirate. The election of an Emir should be by free democratic election within the Emirate and can be by either democratic method outlined below. However, the entire Dar-al-Islam must be run as a Caliphate, otherwise the government of Dar-al-Islam is not following the Sunnah and the following of the Sunnah of the Prophet Mohammed (saw) is not optional.

Creating a Caliphate in today's world is not going to be easy but it is not impossible. There are matters of great importance in my 'Theological Advice', below, that should make it easier. Whilst it may not be or may not seem to be necessary to have an 'Immediate Caliph' appointed until an 'Elected Caliph' is selected by the umma, if it IS felt that there should be someone in charge of the Caliphate once one is declared, then I am of the opinion that the best person for the job is the current reigning monarch of the Jordanian Royal Family, King Abdullah II. There are three reasons for my choice:

1) Religious. The current Royal Family of Jordan proclaims to be Sunni Muslim and, therefore, is acceptable to the majority of Muslims worldwide.

2) Genetic. The members of the current Royal Family of Jordan are direct descendants of Ali's eldest son, Hassan, and therefore, they are, by virtue of their birthright, acceptable to Shi'a. (If they are not deemed acceptable, then it is because of modern political and religious disputes and bigotries and carries no weight with the original intentions of 'The Party of Ali' and those whose eyes and hearts are open to the truth know this to be true.)

3) Political. The current Royal Family of Jordan was, formerly, the ruling family of the Hejaz for over 600 years. That family protected the holy sites of Mecca and Medina for that period and only left the Hejaz during the Arab Revolt, when that family decided to take opposition to the Ottoman Turks. While they were fighting the Ottomans, the Al-Sauds swept westward and overtook the Hejaz in their absence.

When the fighting against the Ottomans ended at the end of World War I, the family could not return to the Hejaz and rule, as the al-Sauds - now lushly enriched with their newly discovered oil reserves - with the backing of the Wahhabis, had usurped the Hejaz and the Haram behind their backs. Tribally, the Al-Sauds had no claim to the Hejaz whatsoever; it was only the wealth and power of their oil and the zeal of the Wahhabis that led them to take that which was not theirs. Nevertheless, what's done is done (and those truly responsible are all, now, dead) and no new war or battle over the Hejaz would be just. So, the Hashemite family settled in the Trans-Jordan and have nobly led that region ever since.

It is my opinion that these three reasons give that family the best claim to the Caliphate and, so, if it is felt by the umma that there should be a de-facto Caliph prior to elections, then this family is, by far, the most suitable to be thrust into that position as they have a religious, a genetic and a political claim with which to be reckoned that is far superior to the claim of any other family worldwide. Furthermore - it can hardly go unstated - that, in this author's opinion, that family's honour, nobility and courage throughout the years has more than proven them to be capable and worthy leaders. However, that does not sway me from believing that the Caliphate should, and of right ought to be democratic and that the Caliph, as Abu Bakr made clear, is answerable to the umma and only rules by the assent of the umma. Therefore, because of the Sunnah of Abu-Bakr, any initial appointment of a Caliph would still need to be ratified by a popular vote of the umma in order to ensure that the Caliph is ruling with the assent of the governed. And, now, I suggest how that process could be carried out - with or without any initial Caliphatic appointment.

Returning to the process of electing a Caliph, each Emirate - an Emirate being a former (and, at the moment, current) 'Nation

State' that declares itself as a Muslim Nation - should put forward a representative who they feel is best suited to be Caliph. This representative may be the current Emir; but, I feel that should be discouraged, as it could lead to what might be conceived as a coup by an Emirate to seize control of the Caliphate, which may bring Dar-al-Islam into disrepute with their non-Muslim neighbours as well as themselves. These representatives should be elected by the populations of their respective Emirate in fair democratic elections - either by a first-past-the-post or proportional representation method. Either method used is completely up to the Emirate, as it may be fairer in certain Emirates to run a first-past-the-post method and in some Emirates, proportional representation; it depends on the internal dynamics of the population of the Emirate. The methodology used should be voted on by the population of the Emirate; as there are only two choices available, neither methodology would impugn the voting for the methodology used, as the method chosen would be chosen by simple majority. Then, once each Emirate has put forward their candidate, a second election - also a fair, democratic election run across Dar-al-Islam - would determine the Caliph. In the Caliphatic election, each Emirate's population can vote for any candidate that has been put forward by any Emirate and, again, those elections can be by either method. Once an Emirate has chosen a winner, the Emirate puts forward their single vote for that candidate and the Caliph is chosen from a simple majority of Emirate votes, which, for the first election only (and in certain cases outlined below), becomes a default first-past-the-post election.

The Caliph would only reign whilst the public support them. This is based on the Caliphate of Abu Bakr, the first Caliph, who declared that he should be treated thusly. If the public withdraw their support, then a new Caliph would be elected. Each Emirate would hold elections every 5 years A.H. in order to establish public support of the Caliph and, the Caliph would retain their seat with a simple majority or a new Caliph would be instated if the majority of

Emirate votes fell to an opposing candidate. In each 5-year election, the Emirates, by the original method, would elect a potential successor. Following that election, an election is held to establish which of the two candidates is Caliph. That way, all subsequent Caliphatic elections would be based on a simple majority of the entire Dar-al-Islam, as there would only be two candidates to vote for and each Emirate has a single vote for the Caliphate; thus, over 50% of Dar-al-Islam would be in support of the Caliph. The only exception, of course, is the original election (and elections where the Caliph is impeached by virtue of acting in poor faith, as described below) and after 5 years, the people can overturn their choice.

If it is felt by the umma that the Caliph (including the first Caliph) has acted in such a way as to bring disrepute to Dar-al-Islam, the Emirates must hold an immediate special election in order to determine the public's opinion and, again, a simple majority of Emirate votes would allow the Caliph to retain their seat. If it is deemed that the Caliph is no longer supported, then the Emirates must, without delay, hold another election to put forward a candidate and the original process that applied to the first election of a Caliph is re-applied. Because this process effectively returns the Caliphate to a position where the Caliph elected is elected via first-past-the-post, it returns the Caliphate to a situation where the Caliph may not be supported by the entire Caliphate; therefore, it is strongly urged that the Caliph always act in such a way as to avoid such public impeachment by Dar-al-Islam, as it weakens the strength of the Caliphate and weakens the support of the Caliphate by the people, as well as in the view of non-Muslim nations.

Once a Caliph is elected, they will appoint a Vizier, who will act as a second-in-command. The Vizier should be an individual chosen for their knowledge, wisdom and piety. The Vizier is then ratified by a popular vote by the Emirates and is given the title by simple majority, i.e., over 50% of Emirate single votes. If the Vizier is not ratified, then a new appointment is made until a Vizier is ratified. The Vizier would act as an advisor to the Caliph and would serve as Caliph in interim situations such as in an impeachment/confidence vote against the Caliph or upon the death of a serving Caliph, thus, the importance of the Vizier's knowledge, wisdom and piety.

Upon the death of a serving Caliph, Dar-al-Islam must, without delay, elect a successor according to the rules above. The Vizier may be a candidate for Caliph if their own Emirate elects them as their candidate; but, the Vizier is not automatically a candidate for Caliph.

I believe it is in the best interests of Islam that the Caliph's political seat is based in Medina, Arabia. This is because that was the first city that was governed by Islam and Shari'a law; plus, it brings full circle the spread of Islam worldwide back to its origin. That does not mean, however, that the Emirate of Arabia holds more power than any other Emirate simply because the Caliph reigns from within their territory. The Caliph represents the entire Muslim umma and is responsible to and for the entire Muslim umma. Medina, unlike Mecca, has not been given any special political privileges, as Mecca has been given religious privileges and responsibilities. It is used as the base for the Caliphate for historical reasons, reasons based on the Sunnah of the Prophet (saw) and because, in today's age, the position of a capitol city is not required to be centrally located or strategically located because modern weaponry can protect (and, unfortunately, attack) any part of the planet.

But who should be eligible to be a Caliph? It best serves the umma that the Caliph should be one who knows the full meaning and content of the Qur'an and has memorised it in Arabic, i.e. a potential candidate for Caliph should be Hafiz. The Caliph should already be a Hajji; otherwise, their hajj is likely to become a political and/or media event and destroys the concept of equality of their hajj. Likewise, any Emir should also be a Hajji, for, if they have enough money to launch a campaign to be elected Emir and have not taken the Hajj, their money would be better spent on taking the Hajj first, as their duty to Allah is a far greater calling - and an obligatory principle of Islam, whereas being an Emir (or wanting to be an Emir) is, by no means, obligatory - than any duty they may feel to lead their fellow Muslims however earnestly that duty is felt. The Caliph should be known to be extremely pious, that is, they pray 5 times a day, pay zakat, and fast during Ramadan and, as stated above, is already a Hajji. They needn't have ever been an Imam, but they should be able to serve in that role if it is ever required; in other words they should know how to lead a congregation in prayer including the Adhan, the call to prayer. The same stringency should also apply to the Vizier for the very same reasons.

As Caliph, they must act in the best interest of Dar-al-Islam; this is their greatest and most solemn duty. Predominately, this is done in order to protect the integrity of the office due to the election process; however, the Caliph has an enormous moral weight on their shoulders, as they represent the entire Muslim umma. As the representative of the umma, in the eyes of both Muslims and non-Muslims alike, their duty to act in the best interests of Dar-al-Islam is the most difficult responsibility a Caliph holds. They should not start wars against non-Muslim nations but are fully free to wage war in the defence of Dar-al-Islam if any part of it is attacked. Thus, the Emirates are bound by an implied treaty where, if any Emirate is attacked, the entire Caliphate will come to their defence. This, however, does not imply that a nuclear strike is ever sanctioned as a means of defending against a conventional attack. Rather, nuclear arms are intended as a deterrent against nuclear attack and can only be used if any part of Dar-al-Islam is first attacked by a non-Muslim nation by nuclear weapons. If, for example, a radical Muslim sect attacks Dar-al-Islam, with any form of nuclear device, from within or attempts to make it look like an outside attack, nuclear response is prohibited; thus, the nuclear option is only executed when the Caliph and Dar-al-Islam is absolutely certain that they have been attacked by nuclear arms, without provocation, by a non-Muslim nation. And, of course, it is prohibited, above, to strike first with nuclear arms, so provocation is impossible, as it is an extremely un-Islamic act, as well as being cowardly and foolish and any first strike with nuclear arms would, immediately, bring Dar-al-Islam into disrepute and the Caliph with it and, of course, bring the full wage of the anger of Allah onto the head of the Caliph.

It is also wise to not start wars for lighter causes than such as being attacked. If a disagreement with a non-Muslim nation arises, then diplomatic solutions are called for. There is no reason to levy war unless there is no other way to resolve the issue and that means that all other means of resolution must be fully explored and exhausted before war is even considered. Islam is a religion of peace and we, as humans, are reminded by the Qur'an that we were created to know one another, not to kill one another. Diplomatic solutions are sometimes more difficult to arrive at than simply declaring war, but it is not the Islamic way - the Treaty of Hudaybiyah should stand as the greatest example of that principle to those who may have forgotten it.

The Caliph should always endeavour to keep peace with the Dar-al-Harb, that is, non-Muslim nations; however, that term itself, Dar-al-Harb, is a poor and unhelpful term, as it means 'the world of war'. It would be better to refer to non-Muslim nations as the Dar-al-Dawa, the world of invitation, who are always invited to join the Dar-al-Islam and every Caliph should, at the beginning of their reign, invite all other nation states to join Dar-al-Islam. This is based on the Prophet's (saw) invitation to neighbouring States to embrace Islam. The Caliph makes these invitations, then, as it is Sunnah and non-Muslim nations should not be offended by this practice but understand it for what it is - that it is a religious obligation to follow the traditions of the Holy Prophet (saw) and, as the Holy Prophet (saw) invited all other nations to join Islam, this invitation is done in remembrance of that.

Of course, there may be a time when a non-Muslim nation decides that it is politically advantageous to join the Caliphate. That nation will be reminded that its citizenry will be subject to Shari'a law after joining and, therefore, any nation wanting to join the Caliphate should hold an internal election to see if its own citizenry are in favour of joining Dar-al-Islam. The joining nation cannot and will not be forced to become Muslims nor will its citizens be forced to convert, as it is stated clearly in the Qur'an that there can be no compulsion with religion; however, they will be subject to Shari'a law and, most likely, the tax that is associated with non-Muslims living in a Muslim nation. The citizens of a joining nation will be reminded that the tax is levied as opposed to the enforced and religiously obligatory payment of zakat and, therefore, they are not being treated unfairly financially.

At the beginning of each year A.H., there should be a council of Emirs with the Caliph and Vizier in Medina to establish policies for Dar-al-Islam. This council should last 10 days, thus ending on the 10th of Muharram, Ashura, and the Emirs, Caliph and Vizier are, on this day, encouraged to fast during their last day of council. This makes the council both a political event and a pious event. As the world is ever-changing, an annual council should make it easier to help Dar-al-Islam react to the necessities of Dar-al-Islam and the necessities of the world at large. After that council, the Caliph should declare the decisions of the council to the United Nations, so that there is transparency between Dar-al-Islam and the Dar-al-Dawa. This will help ease relations worldwide and ensure that the Dar-al-Dawa knows what Dar-al-Islam's policies are. Also, there may come times of international crisis and a session with the Council of Emirs may be called by the Caliph in order that Dar-al-Islam can react to the crisis. This crisis may be political, such as war breaking out between any nations, or the crisis may be humanitarian in nature, and, as such, provides Dar-al-Islam with an opportunity to demonstrate its charity. In any case, the Caliph has the power to call an assembly of the Council of Emirs as he sees fit; however, this should not be done lightly, as it incurs certain expenses (the travel costs, accommodation and, of course, the fact that it demands that the Emirs leave their own Emirates with their second-in-commands in charge and, in so doing, the Emirs are not present in their Emirate to react to crises at home should they arise. Also, there are the inherent security risks involved.), and the power to call an emergency assembly should not be flouted, as this will bring the Caliph into disrepute.

As the Qur'an states that the People of the Book are protected people (Dhimmu), Dar-al-Islam can make no aggression against the Jewish State of Israel. This must be a unilateral decision by Dar-al-Islam to not act against Jews based purely on the dictates of the Qur'an. Anything less is un-Islamic. Considering the Palestinian question, Dar-al-Islam must use diplomatic means to resolve the issue and must, if needs be, lay down any arms against Israel. That way, any further aggression could only be by Israel and would only serve to bring Israel into disrepute in the eyes of the world at large. This is, in a sense, a modern-day Treaty of Hudaybiyah, where, at first, it appears that Dar-al-Islam is weakening itself against a supposed enemy; however, the Jews are NOT an enemy to Islam (this concept will be explained further in the 'Theological Advice', below) and such a unilateral action implies the faith of Dar-al-Islam in Allah and will pave the way for a peaceful resolution, which is what both sides should desire in the first place. Likewise, Dar-al-Islam should not and, by the same Qur'anic principle, cannot make aggression against any 'proclaimed' Christian nation. Nor should Dar-al-Islam make any aggressive statements against Christian or Jewish beliefs or customs, as this only engenders mistrust and is antithetical to Islamic principles.

With respect to non-Jewish and non-Christian nations, Dar-al-Islam should not act in any way as to antagonise them either, for

271

there is no Islamic principle that calls for that. Rather, it is Islamic to work together with non-Muslims for peace worldwide and for mutual benefit. Whilst reason alone should dictate that it is politically imprudent to antagonise a neighbouring nation simply because their citizenry hold a different religious view, it falls under the guiding, Qur'anic principle of Islam that 'people were created to know one another' and for no other reason. Therefore, Dar-al-Islam should endeavour to work with non-Muslim nations in order to better know them and that Muslims can interact with non-Muslims and demonstrate the kindness, charity and egalitarianism of Islam. This would serve to aide international relationships and, at best, may serve as the basis for non-Muslim nations wanting to accept the invitation extended to them to join Dar-Al-Islam and, God willing, Islam itself.

I firmly believe that it was the dream of the Prophet (saw) to envision an Islamic world. This may not be possible today but it can always be a goal of the Caliphate guided by the principle that there can be no compulsion with respect to religion. Imagine a world where there is a nation that stretches from Morocco to Indonesia: one people, one faith, one God. If there is any concept in Islam regarding a trinity it is those three: one people, one faith, one God. This Dar-al Islam would be a nuclear power, via Pakistan, and be a member of NATO, via Turkey. There would have to be a permanent place for it in the UN's Security Council. Finally, Dar-al-Islam would be taken seriously and, I believe, the world would be a safer and better place for it.

... Theological

There needs to be a resurgence of Ijtihad (interpretation of the Qur'an) in the modern age. My argument for this is as follows: The ulema of the past have closed the 'Gates of Ijtihad' because they felt that all interpretation had already been done and no further interpretation would be required. However, they hadn't foreseen the great changes that have taken place in the world since their time, especially with respect to our scientific understanding of the universe, which is only a better understanding of the details of how Allah created and/or maintains this universe. First, we must assume that the 'Gates of Ijtihad' were opened, if we can accept that any Ijtihad occurred in the past. As we know it did, we know that the gates were, at some point in the past, opened. Now, if they were opened by humans and were closed by humans, then humans can re-open them at any time; however, if the gates were opened by Allah, then no human could EVER close them, as no human has power over that which is omnipotent.

So, it can be logically determined that Ijtihad can take place in the modern world and, in the face of certain scientific findings, including those put forward in earlier chapters of this book, there should be some new interpretations, some of which, of course, I've already alluded to but today's ulema should also be encouraged to make new interpretations so that Shari'a can be updated to address modern issues backed up by modern knowledge. It is only reasonable that wisdom - the determined action based on knowledge and understanding - as envisaged by, encapsulated in and executed by Shari'a law, be continually updated to reflect new knowledge and understanding of the universe.

Islam accepts that Allah is One and shares that view with Judaism and with Advaita Vedanta Hinduism, assuming an equality between YHVH, Brahman and Allah and, as each of these faiths believe in only one God, logically, it MUST be the same, irrespective of the name used to denote that entity. Islam acknowledges Muslims as siblings of the faith, thus calling one another brother or sister; however, Islam also acknowledges that all humans are part of a larger family - the family of Adam. Because of that, Muslims should be encouraged to refer to non-Muslims as 'cousins' like they refer to one another as brothers and sisters. In doing this, it will re-enforce that feeling of a familial bond between all humanity and re-affirms their core understanding of the principles of faith with respect towards acting towards any human laid down in the Qur'an. It should also engender less enmity between Muslims and non-Muslims and encourage a greater trust of Muslims by non-Muslims. This should act as a basis to sweep away some of the distrust that has accumulated recently and, perhaps, lay down a strong premiss for non-Muslims to accept the dawa and become a part of the Muslim faith (Islam) and polity (Dar-al-Islam).

For example, in modern England, it is common usage for a Muslim shopkeeper to refer to a non-Muslim customer as 'Boss'. This is similar to the old 'Raj' usage of the term 'Sahib' and implies a master/slave or superior/inferior relationship where the shopkeeper self-deprecates in order to be well-perceived by the customer. The usage of 'cousin', however, implies a more egalitarian stance and puts the two on a more equal footing with one another and that is more in keeping with the Islamic view that no one is more important than any other except for reasons of piety, which, in the case of a customer/shopkeeper relationship, could simply never be known - except by Allah.

A major problem in today's Islam is the diversity of sects. The Prophet (saw) warned of the dangers of sects and declared that only HIS sect would be considered true Muslims on the Last Day and the Prophet (saw) only ever declared himself to be a Muslim. So, whilst tolerating some diversity in practice, no Muslims should ever refer to themselves as a member of any sect. That means no longer saying, "I'm Sunni" or "I'm Shi'ite" or any number of other sectarian names. Rather, Muslims should only state that they are Muslims and, if asked, "What kind of Muslim?" should respond with, "a faithful Muslim" and a smile. The term 'Sunni' is implied by the term Muslim and, so, its usage is redundant at best and, at worst, horribly divisive. Equally, however strongly one feels about the events at Karbala and the real and/or perceived injustices done to the family of Ali, to refer to oneself as

"Shi'a" is, again, divisive with respect to the umma. Allah permitted those events to occur for reasons known only to Allah - perhaps so that, now, we can drop the appellations of sectarian terms in order to return to a more faithful Islam and, finally, BEFORE the Last Day, correct the problem of sects within Islam, as the Prophet (saw) would have us do. Remembering those events is perfectly fine and learning from them even better, but to persist in making them an excuse to divide the umma is fruitless and un-Islamic.

I have mentioned to many people the dropping of the term 'Sunni', who use that term to refer to themselves, and have had mixed reactions, but the more learned the individual is in Islam, the less of a problem they have. The more learned the listener is, the more they know I'm right. Thus, a strong Islamic education throughout the umma with respect to the problem of sects and the fears that the Prophet (saw) had regarding them, should serve to alleviate any resistance to dropping them. In short, if there is an adjective in front of the term Muslim - other than 'faithful' - then it is an adjective that only serves to divide the umma against itself and the Prophet (saw) warned against that.

Muslims must remember that the People of the Book, Jews and Christians, are protected people and must, at all costs, avoid confrontation with them. Arguing and debating theology is, of course, perfectly fine, but not throwing stones or assaulting them just for believing differently or even retaliating violently if they levy an insult at you. If they insult Islam or the Prophet (saw), then it is better to just walk away and know that Allah will treat them justly. This would be the policy of Isa (Jesus) and it is encouraged to remind a Christian of that before walking away. If it is a Jew who insults Islam or the Prophet (saw), then remind them that their religion forbids slander (lashon hora) and that they have broken a precept/commandment of the Torah, and then walk away. Islam is a peaceful religion and the only way to demonstrate that to others who try to rile you to aggression is to fight with strong words that prove that you know more about their faith than they themselves do.

Muslims should invite Christians and Jews to Islam but never in an aggressive or overbearing manner; however, not all Christians or Jews will accept the invitation and there is no need to insult their religious sensibilities. For just as faithful Muslims would turn down an invitation to Judaism or Christianity, faithful Jews and Christians would turn down the invitation to Islam for the same reasons. If they refuse the invitation, then Muslims should encourage them to more strongly follow their own faith. With respect to the Jews, this would mean encouraging them to rebuild their lost Ark of the Covenant - even going so far as to help them to do it - for Muslims have nothing to fear from the Ark of the Covenant. I speak more about this, above, in the advice to Judaism. With respect to Christians, again, Muslims should encourage them to discover more about the roots behind Isa's (Jesus') true faith and remind them that Pauline doctrine was never a part of Jesus' teachings.

Regarding Hindus, Muslims should encourage those who are Dvaita (dualistic or polytheistic) to look back to their roots and rediscover the Monotheism/Monism of the more ancient Advaita Vedanta approach, which was the original Hindu faith. Whilst I develop this further in my advice to Hinduism itself, if a Hindu worships Brahman, then no Muslim could ever have a problem with that, as Brahman is an all-encompassing single deity that is completely cognate with Allah and a worshipper of Brahman is not polytheistic in any way. Although their practices differ, this is, no doubt, because the prophet who brought Advaita Vedanta to them was of a different cultural background to the Semitic culture from which sprang the three Abrahamic faiths. This prophet was, probably, of a branch of Noah's son Japeth who had moved east, as the Aryan peoples are an Indo-European people of the Indic branch, they must be Japhethite in origin.

To return to the Jewish people for a moment, if they do rebuild or find their lost Ark of the Covenant, then they would want to rebuild a sanctuary for it. They should be reminded that the first sanctuary was at Shiloh and that the Torah never states that the Sanctuary is required to be on the Temple Mount. The Torah only stipulates that the Ark be kept in a Sanctuary; it never states where that Sanctuary needs to be. Although, there is a strong feeling amongst modern Jews that it should be rebuilt on the Temple Mount. If the Jews persist in this, then remind them that there is enough space between the two structures that currently exist on the Temple Mount, The Al-Aqsa Mosque and The Dome of the Rock, and that a Sanctuary/Temple could be built between those structures and there is no need to destroy the main structures that are already there.   This is the essence of compromise.   Remind them that The Dome of the Rock is, essentially, a sacred building that sanctifies the place where they believe Ibrahim (Abraham) was willing to sacrifice his child and that the Al-Aqsa Mosque is also an edifice built to worship the One true God, which must be the same as YHVH, the Jewish name for Allah. To destroy either of these structures would be against the tenets of Judaism, as both of the existing buildings are dedicated to the worship of the same, single deity.

The Temple Mount is sacred to both Judaism and Islam and should be protected by both faiths together. The Christians, too, because their faith grew out of Judaism, also hold the Temple Mount to be sacred, although they feel less strongly than the Jews about it. However, the simple fact remains that there is ample space between The Dome of the Rock and The Al-Aqsa Mosque for another structure and Muslims and Jews alike should bear in mind that they worship the same God, albeit in slightly different ways because they were each given their own revelation; the spirit of compromise between these 'brother-faiths' should be encouraged in order to lay down a foundation of trust between them.

Ishaq (Isaac), the patriarchal father of the Jews, and Isma'il (Ishmael), the patriarchal father of Arabs and Muslims, died as friends and loved each other as the brothers they were; they would be horrified at how their descendants have treated one another - especially recently within the past century. So, honour their memory and strive towards reconciliation with Jews wherever possible. Allah has chosen both to be guardians of their respective revelations and chosen both to stand as examples to the rest of the world. It's high time that the followers of these faiths take that to heart and act upon it.

Regarding the A.H. calendar system, Islam has always determined the start of each month by visibly observing the new moon; however, it is often seen a day (or, sometimes, two days) late and this causes some months which should have 29 days to have 30 and vice versa for compensation later. The Qur'an states to 'observe the New Moon' but does not say 'visibly' or 'with one's eyes'. This would allow - or should allow - that modern scientific techniques of observation are perfectly valid 'observations'. If scientific observations were used, there would be no discrepancies ever. The planetary formulae are well known in today's world and can serve far more accurately than any visible observation ever could. Islam is the one Abrahamic faith that has never discouraged scientific discoveries, as it understands science as a fine method of discovering the intricacies of Allah's creation. Islam should welcome, with open arms, the usage of science in any arena and the calendar system should be no exception. Almost every year at Ramadan, the world makes fun of Muslims as some Muslims begin the fast a day before others. Scientific observation would remove that possibility and let the world know that Islam is a modern religion, albeit with strong, historical roots like Christianity and Judaism. No longer would these faiths have the chance to poke fun at Islam as a 'backward' faith that is hopelessly lost in the past; rather, it would show them that Islam has come of age and is willing to modernise where modernisation makes sense. And, in this case, it does make sense, as it brings the entire umma together and brings together the timings of their worship, especially with respect to Ramadan and both Eids.

The best argument in favour of retaining visual sightings of the lunar crescent is that it maintains a link to the natural process and this is a strong argument, as well, because it keeps Muslims in a state of awareness of God's Creation. Nevertheless, this same feeling can be garnered from the precise calculation of the new Moon. But unity of the practice of Islam can only be gained by moving towards more modern observation techniques.

There has been a history of allowing certain cultural practices to continue to be practiced when those practices are, in essence, un-Islamic - in particular, female circumcision. Female circumcision must be abolished throughout Dar-al-Islam, as it is a practice of mutilation that the Prophet (saw) would never have sanctioned and there is no basis for it in the Qur'an either. It was allowed for expedience in order to gain followers and was allowed as a compromise to those who practiced it. However, the time has come for the complete abolition of this rite and a fatwa should be declared and Shari'a law be amended to enforce this. Female circumcision, which is the removal of female prepuce, labiectomy and clitoridectomy (either partial or radical), has never been an Islamic practice - it is a cultural practice - and education is required in those places, mostly in Africa, in order to educate them against it. There will be staunch opposition to this, as many males in these cultures are brought up to not want a woman who has not had these procedures performed on them; however, when presented with no alternative, they would soon change their minds, as I'm sure the instinct to procreate is stronger than any cultural adherence to such abominable practices. The egalitarian aspect of Islam must reach out and protect these women who have been abused for centuries and stamp out, forever, this horrid practice. To not prohibit this practice is to forsake hundreds of thousands of female members of the umma and it is long overdue, as its prohibition is essential to the overall integration of the entire umma. This is another subject where non-Muslims, at the moment, have the moral high ground when they stand against it. Islam should have the backbone to listen and be admonished when non-Muslims are correct in their admonishment and to take corrective measures against a practice that, whilst practiced by a minority of Muslims - like terrorism - nevertheless brings all of Islam into disrepute.

With respect to punishment for female circumcision, I suggest that it be viewed as theft and a form in which the property can never be returned. Further, it is a theft of dignity, as most girls who have this done to them are forced and physically held down during the procedure, which is usually done without anaesthetics and, often, in an unsanitary environment that can lead to infection and/or death. Therefore, cutting off the hands (both hands!) of those who perpetrate it is my suggestion for punishing those who commit this barbaric crime; furthermore, anyone found guilty of conspiring in a female circumcision should also have their hands cut off. The punishment is made severe in order to deter the crime and demonstrate the abhorrence with which Islam holds this practice. If only a fine were imposed, then people would simply pay the fine and continue the practice. In order to abolish it altogether, an extremely strict punishment is required even to include not only the person who performs the operation but to include anyone who has aided in the process. This would include both parents, if they conspired to have their daughter circumcised, and anyone who held down the girl in order for the operation to be performed and, perhaps, anyone who sought out a physician who would be willing to perform the operation. Only by setting forth a strict punishment - and, when encountering the crime, enforcing the legislation to the full extent - can Islam be assured of putting an end to this practice and, in so doing, gain respect worldwide for finally taking a staunch stand against this abominable and most un-Islamic practice.

Male circumcision, on the other hand, seems to have become an optional aspect of Islam in many places and this, too is wrong, as it has always been the physical mark on the body of a male of their acceptance of the Abrahamic covenant. For Islam to turn its back on this foundational aspect of Abrahamic tradition is to break that Abrahamic covenant. Islam prides itself on being 'the Cult of Abraham', if so, then it must enforce male circumcision or risk being hypocritical with respect to the Abrahamic covenant. This may result in fewer converts, but those who do convert would be those whose faith was, obviously, stronger; so it ensures the quality of the faith of converts and deters conversion for the sake of societal acceptance. In today's world, the operation can be performed completely safely and with extremely little risk of infection. The individual is usually fully recovered in 3 weeks. Plus, it has been demonstrated that circumcised males are less likely to succumb to venereal infections and are less likely to spread them, as well. These facts should be impressed on any prospective convert in order to allay their fears regarding it.

Also, regarding the circumcision of Muslim boys, it is a tradition to put it off until the child reaches 13, in memory of Isma'il's age at the time of his circumcision. Whilst there is nothing necessarily wrong with this, it is not required to honour this timing, as it was a happenstance with respect to Isma'il. The covenant of circumcision came to Ibrahim when Isma'il was 13, but the covenant came to Ibrahim, not Isma'il. It would be better to do it at a much earlier age, like the Jews, who perform it 8 days after the child's birth, after the manner of Ishaq's circumcision. Of course, I'm not advocating following the Jewish pattern; however, it makes it far easier on the child to do it nearer the time of birth. Perhaps a compromise between the two traditions could be applied. Rather than wait until the child is 13 years old, wait until the child is 13 days old and use the day-for-a-year method to honour the timing of Isma'il's circumcision. Remember: Islam is not a cruel faith and should always strive to make easier those things which are hard. The problem still remains, of course, for converts, who must be circumcised at whatever age they proclaim Shahada, if they are not already circumcised. And this, I'm afraid, will always be hard but it is not so difficult that it should be a barrier for a person who truly accepts Islam, as it is not a barrier for converts to Judaism, as both Orthodox Judaism and Conservative/Masorti sects are strict when it comes to this aspect of conversion to Judaism. Rather, it becomes a marker of the extent of the faith of the convert and proves that the convert is earnest in their acceptance.

Those groups that have employed terrorist tactics like car bombs and suicide bombings ought to be formally ostracised by the umma. These acts are completely un-Islamic and have given the worst possible name to Islam that it has ever had throughout its history. In order to regain the trust of the world's population, the mainstream umma must decry these tactics formally and there should be a fatwa declaring that any such persons or groups who employ such tactics are not Muslim but are guilty of apostasy (Irtidad) and are no longer to be considered Muslim in any way, shape or form. The international declaration of this fatwa is imperative at present and ought to be agreed by every nation that considers themselves Islamic; however - if it has not been declared before a Caliphate is formed - it is certainly a priority of the Caliphate to either maintain that fatwa or to declare it as soon as possible. To use car bombs that kill indiscriminately is not in keeping with Shari'a law. To convince an individual to strap a bomb to themselves and set it off, killing both themselves and, indiscriminately, others around them is not in keeping with Shari'a law. These methodologies are certain gateways to Jahannam (Hell) and not Jannam (Heaven), as their leaders would have them believe. To call them martyrs is, again, a fallacy; rather, they are cowards who haven't the courage to sit at a table and discuss terms with their enemies or to fight them honourably. These are not methods of (lesser) Jihad, they are the methods of cowards and apostates and the mainstream umma must stand up as one and denounce them and the ulema must publicly declare that anyone who uses these methods or condones them are guilty of Irtidad (treason against Islam) and that the umma no longer recognises them as Muslims. Most of the non-Muslim world currently views Islam as a religion of terror because of these few people and complacency by the umma at large has, in non-Muslim minds, verified that. In order to turn that around, Islam must clean up its own act and show non-Muslims that true Muslims abhor such acts and that they will no longer stand idly by and allow these treasonous elements to rob Islam of its good name. To be fair, in many cases, it is not complacency but fear of the terrorists themselves that prevents a nation or its people from decrying these acts; however, I remind you that "you should only fear Allah" and that fearing these terrorists is, in its own way, turning your back on this principle of Islam.

The lesser Jihad (going to war) is either just and warranted or unjust and unwarranted - it is never 'Holy'; so, if someone states that they are in a 'Holy War', they must be rebuked, for mankind was not created to fight one another but to know one another. If a war is proclaimed to be just, then it is to be fought honourably against a known enemy. No one is allowed to simply attack a marketplace in the hope that some people there may actually be combatants. That is, surely, overstepping the bounds and Allah could never be pleased with such an act. The usage of car bombs and suicide bombers - especially by those who proclaim themselves to be Muslim - is an outrage against Islam and an attack against Islam, even if - and especially if - the perpetrators do claim to be Muslim. The very act proves that they are not. Therefore, it should be incumbent upon Muslims to uncover such plots if they can and prevent them if they can and, if it is at all possible, the conspirators and/or perpetrators should be punished to the fullest extent of Shari'a law (in other words, they should be put to death) in order to regain respect in the eyes of non-Muslims, fellow Muslims and, most importantly, in the metaphorical eyes of Allah.

... The Palestinian Problem

The question of Palestine requires a third section as it is both a political and theological situation. There is a solution but it will be difficult and will call on Muslims worldwide to show restraint if they are faced with aggression. Addressing this problem would take the fuel of the fire of anger away from those extremists who use it as an excuse for terrorism against nations whom they believe support Israel and allow Israel, by way of ignoring the oppression, to continue their institutionalised aggression towards Palestinians, who are, in fact, a brother race to the Israelis. This conflagration is a fraternal war that must be stopped, as the respective fathers of these peoples, Ishaq (Isaac) and Isma'il (Ishmael) would be ashamed and horrified to know that their descendants are killing one another on a daily basis over land which is sacred to both of them.

The situation is largely a result of the failure of the Arab Palestinians to agree to the original partitioning of Palestine; however, that partitioning would never have worked anyway, so it is not surprising that the Arab Palestinians refused to accept it, as it was doomed to fail. Since Israel agreed and the Arab population refused, Israel set borders, declared their State and then began work to reclaim the areas not formally accepted as Arab/Palestinian as their own because there was no 'Palestinian State' to stop them. This, in turn, has caused a sequence of battles and wars over the land. The recent solution of having two separate areas for Palestinians, the West Bank and the Gaza Strip, is not working for the fundamental reason that it has, historically, always been difficult if not impossible to govern a country that

284

does not have contiguous borders. East and West Pakistan is a clear example of this within a strictly Muslim population and led to the creation of an independent Bangladesh from the former East Pakistan, as a means of rectifying the ineffective partition.

So, recognising that the greatest barrier to a successful Palestine is the failure of the condition of partition, naturally leads to a solution that eradicates any partition and re-instates a Palestine that has contiguous borders. Also, it seems obvious from the continual insurgence into the Gaza Strip that Israel is intent on governing that land by virtue of displacing the current residents and re-housing their own growing population in that region. At the same time, Israel seems to not be that concerned with the area just south of the West Bank, which borders on the northern regions of the Negev Desert.

I suggest, then, that the West Bank area be increased to the south by an area equal in area with the Gaza Strip and that this new, enlarged West Bank be considered Palestine. It would lose the ports on the Mediterranean and any advantages of having coastal area; however, this is a small sacrifice to make for lasting peace and sovereignty. Again, one must remember the Treaty of Hudaybiyah and recount that it is sometimes worth making seemingly great sacrifices for a greater good; to remember this in the spirit of the Treaty of Hudaybiyah makes this act an act of following the Prophet's (saw) Sunnah. As I've mentioned, Israel seems intent on taking the Gaza Strip by force, so it is easier to just let them have it. Fewer lives are lost and so many lives have been lost over such a small amount of land that it is reasonable, peaceful and Islamic to forego that land for a peaceful solution. That way, those who have died will have died for what, ultimately, became a struggle for peace and their deaths would not have been in vain and would have an everlasting, redeeming value.

So, while Palestine loses the Gaza Strip, it gains an equal amount of land south of the West Bank. Equally, Israel loses an area south of the West Bank and gains the Gaza Strip, so both sides

285

gain something and lose something; this is the essence of compromise. The biggest problem here is that the land south of the West Bank is not particularly arable nor are there large housing estates there. This lends both the Arab Palestinians and the Israelis an opportunity to work together to make this area liveable. I suggest that a certain amount of arms from both sides be melted down and reused to make agricultural tools and/or building tools. This would be a symbolic gesture that would, in fact, make the ancient prophecy that 'their swords shall be made into ploughshares' a reality and would, on a spiritual level, help to make the partnership work, as both peoples would be making an ancient prophecy come true.

This land needs to be made into arable land, so that the Arab Palestinians can grow food there. Water from the Jordan River could be re-routed to be used to irrigate the land and there must be a promise - guaranteed by a treaty - that Israel will not interfere with the flow of water downstream into this area. In order to interfere with the water, Israel would have to cut off the flow to the entire West Bank and that would be easily considered an act of war, so the entire world community would, at the level of the United Nations, have to agree to impose sanctions against Israel if they ever tried to act in this manner. Also, there would need to be many houses built in order to house the Arab Palestinian refugees from the Gaza Strip and, therefore, a ceasefire in the Gaza is imperative and must occur before work on this new land begins. Israel must cease its insurgence into the Gaza Strip and allow the people who live there to remain until new homes are established for them in the new area. Once those houses have been built, the Arab Palestinians from the Gaza Strip can move peacefully and safely to their new homes in the 'Southern West Bank'; and, once the Gaza Strip is depopulated, Israel is free to take it over. There should be no despoiling of the Gaza Strip by Arab Palestinians who leave that area, as such an act is un-Islamic and, since a peaceful solution has been agreed upon, such an act is in the worst possible taste and would bring worldwide contempt on the perpetrators of such dishonourable acts.

If the aforementioned set of actions would be agreed upon by both the Arab Palestinians and the State of Israel, then they could both live together in peace. Not only could they live together in peace but they would make an ancient prophecy come true and that prophecy relates to Jewish eschatology, that is, it is one of the signs of the end times. It is a prophecy regarding the everlasting peace that follows Armageddon and, if peace results from this, it could be interpreted as fulfilling another prophecy: that 'the lion shall lie down with the lamb'. Humanity has a real opportunity, here, to make these ancient prophecies come true and to bring an ever-lasting peace to the Middle East. Please, give this plan consideration, as I feel it is the best way forward for both sides. While each side gives up some land, each side gains an equal amount of land; so, overall there is no loss for either side. This is a worthy compromise, a peaceful compromise, a Muslim compromise and a Jewish compromise. It is an act that would make Abraham (Ibrahim) proud of both of his children and the entire world would be able to rejoice and proclaim that these people are just and righteous people. Please, for God's sake, think about it!

Hinduism

Back to basics is what is required of Hinduism. The root of Dvaita (dualist) thought is Advaita (non-dualist) thought. Hinduism needs to reach back and recover its Advaita roots. If my theories are correct, then the teachers who first taught Advaita were absolutely correct and, in truth, there is only Brahman. Brahman is the one thing that encompasses everything and this, in every way, exactly matches the deity that springs forth from this book. There is only one and that one has many names; Brahman, I firmly believe, is one of those names. Brahman is, if there can be such a concept, the 'hero' of the Upanishads. The various texts that form the Upanishads are the kernel of the old faith and will serve as the kernel of any renewed interest in Advaita Vedanta. Then, of course, the Vedas themselves will serve to add detail; but remember that the allegories and the mentioning of other deities are just aspects of the one: Brahman. The other god-names that are aspects of Brahman are similar to the attributes of Allah, e.g. Allah as 'Rabb ul-Alamin', meaning 'the Cherisher and Sustainer of the Worlds', is remarkably similar to the concept of Vishnu as 'The Redeemer'.

A return to Advaita Vedanta will bring unity to Hinduism and will remove any problems of sectarianism within Hinduism, as well. Plus, it will remove tension with neighbouring faiths that are monotheistic like Islam and Sikhism. The writings of Shankara can also add a practicality to the faith and will help those who have the Dvaita views be able to see how they developed from the Advaita. The Brahma Sutra Bhasya is, without doubt, Shankara's greatest work; but it is not an easy read. To comprehend it can take years but those years will reap wonderful rewards upon the soul who takes on that challenge. I should confess that I am still working my way through it. That book shows the elaborate reasoning that Shankara employed to his arguments and, in truth, he is not far from saying the same things as Spinoza. Both of them easily saw the transcendence of God but had trouble seeing the immanence. String theory allows the immanence to be just as easily seen and, I believe if Shankara and Spinoza had access to string theory, they would have immediately seen the immanence with which they both had problems discovering.

The Trimurti of Brahma, Shiva and Vishnu must be seen in a new light. They are only temporal aspects of Brahman and here is how it works. Firstly, understand that each of these three is only a face of Brahman - they are roles that Brahman plays, as Brahman plays all roles. Brahma creates the present, always. Immediately, Shiva destroys the present by moving it into the past and, at the same time, Vishnu redeems the present by moving everything into the future. The Trimurti is simply a way to describe the passage of time.

Is Ganesha really lucky or is it that those who worship and pray for his help are more likely to see luck when it arrives? I think the worship makes a change in the worshipper and heightens their ability to see how lucky they are. But there are still countless people who have walked out of a temple after just having asked Ganesha for help and they are run down by a bus or accidentally killed in some fashion. In these cases, Ganesha didn't remove obstacles; he placed them before the worshipper. But remember who his parents are: Shiva and Parvati; and Parvati is Kali, just in a good mood.

Durga is, simply put, a personification of the power of energy. There is no need, though, to personify - to anthropomorphise - it is better to accept the power of raw energy as just that. It is the 'stuff' of which Brahman is made and it can be formed into anything and so, it can create anything; thus Durga becomes a 'Mother Goddess' because energy is used to create and creation, for humans, is the role of the mother. Durga is called 'The Invincible One' because energy is neither created nor destroyed; anything that cannot be destroyed is, by definition, invincible.

Durga is also called 'The Unfathomable One' because she destroys ignorance. Well, the more we learn about energy, the more we understand about Brahman. Until we fully understand all that there is to know about energy and what it can do, energy remains unfathomable.

Hanuman is known for being able to change his size at will. This is, again, a personification or anthropomorphising the geometrical concept of dimensions. Dimensions lengthen and shorten and, because of that, all the lengths between infinity and the infinitesimal are defined. But why anthropomorphise this? Understand it for what it is; it is the power inherent in geometry and it is incredibly powerful. Brahman is a master of geometry and so, Hanuman is Brahman's - and everyone's - natural servant. The story of Hanuman swallowing the sun is simply a demonstration of the geometry involved in a solar eclipse. Hanuman was able to fly to the Himalaya to fetch herbs for healing Rama; this was a way of describing how, through knowing the ways of folding dimensions, any trip can be made faster by travelling in ways of which others are unaware. Hanuman knows, well, rather IS those very dimensions and there is no need to personify this, as personification and anthropomorphising God only leads to a misunderstanding of the true underlying nature of God and it is the truth of the oneness of the nature of Brahman that makes Brahman the exemplary.

Lakshmi, whose fickleness gave her the epithet of 'The Restless One', is another personification; this is the personification of unpredictability. It is because Brahman is unpredictable that Lakshmi is even noticed. In today's scientific jargon, Heisenberg's 'Uncertainty Principle' is the new way of referring to the latent essence of Lakshmi. It was when I understood that quantum mechanics rests its feet on the still ground of the uncertainty principle is when I discovered that quantum mechanics has its feet in quicksand. The one thing it is absolutely positive about is uncertainty. The goddess who would cut off her own breasts to

290

please the god of destruction, Shiva, is, without doubt, unpredictability personified. Again, though, there is no reason to anthropomorphise unpredictability; accept it for what it is - an aspect of the one Brahman that makes life interesting. Imagine how dull life would be if it were perfectly predictable. No, don't. Don't waste precious time imagining life being different than it is; rather, take part in the life that Brahman – God - has given you.

Sarasvati is another anthropomorphism; a personification of the power of vibration. She is the goddess of speech and the power behind any mantra. This power is, though, an aspect of the one true God, as I've mentioned in the chapter entitled, "How is God's Word Used to Create?" Yet, again, there is no need to personify vibrations. They are how strings of energy appear to act as different particles and form the material basis of all things. Not only the material basis, but, in all likelihood, the vibrations in the spiritual realms serve to act as the basis for all things spiritual as well. Our very souls rest on this power of vibration as well as the spiritual organ we call our mind. Vibration is what allows a guitar to make music; why anthropomorphise that which we know has nothing to do with the human form? It serves only to confuse us regarding the facts. Sarasvati personifies this concept of vibration perfectly; but, to personify it is to miss the point that it is not a personality but an aspect of nature that is completely impersonal. Vibrations are required for all forms of existence known but most of 'that which exists' is not anthropomorphic; so, it is misleading to demonstrate it as being anthropomorphic.

The tendency to personify and deify aspects of Brahman is what needs to be eradicated from modern Hinduism. It detracts from accepting the true nature of reality. Whilst it is true to say that all things anthropomorphic are natural in one sense it is equally true to say that most aspects of nature are not, in the least, anthropomorphic. And Brahman is not like a man. Humans are like Brahman in that they are intelligent entities, but that's where the likeness and image begins and ends. I believe that Hinduism needs to shed these more anthropomorphic outer images and forge forward with the understanding that nature, in its most part, is not personifiable. Then it will be a modern faith steeped in truth and still based upon its ancient heritage of Advaita Vedanta. Advaita Vedanta was one of the first great steps towards a monistic understanding of the universe and it was way ahead of its time. The Dvaita views that have sprung from it have taken it steps away from the raw truth it had and I believe that it is time to take a few steps back in order to take several steps forward.

I can hear the question arise, "What about Moksha?" That is a very good question, indeed. How can one escape from a universe where there is a single omnipresent entity? The answer is: you can't. But moksha isn't about escaping from the universe that Brahman created and is; it is about removing one's soul from the repetitions of reincarnation. As I've stated above, reincarnation is possible and, because it is possible, it must occur to some people. At the same time, it must also be possible to escape from a cycle of forced incarnations - but not without permission from the Creator. So, how does an individual gain this permission? There are several ways. One way is to ask. One way is to learn about the truth of the wheel of Samsara (the repetitions of reincarnations) and to prise oneself from that wheel by understanding nature as it is and, through that, proving to Brahman that you have learned the final lesson and that further incarnations are no longer required.

Another way is to realise that you are not 'bound' to the wheel in the first place. Any binding of oneself to the wheel of Samsara that an individual perceives is simply another aspect of Maya; it is an illusion. Our eyes are easily fooled and our minds are easily tricked. The truth is that we each have eternal existence. No given incarnation is, by necessity, required. To realise that you are, simply, an extension of the one thing that really exists; this is the key realisation: Atman IS Brahman. This could be done by a soul prior to incarnation, which is why I say that no given incarnation is, by necessity, required; however, everyone that has already incarnated is necessary in that incarnation and THAT is a subtle distinction. This is, of course, not new news but the core teaching of Advaita Vedanta. But it is not an easy realisation to make. I sincerely hope, though, that this book will serve to make it that much easier for many to make, as it lays down the physics that show the truth that Atman really is Brahman by extension.

Advice to Individuals

This is the hardest section of all to write. How do I give advice to every single person on the planet? Do I have the right? Do I have the best advice to give? How can I be sure? I can only let my fingers type what my heart feels. I wish I could hold each of you in my arms, comfort you and tell you everything will be alright in the end. But that isn't the case. If my theory is correct, then this entire universe will end in a cataclysm that no one can prevent. For that very reason, no one can be held responsible for that event and no one can be held accountable for trying and failing to protect us from it because they would fail. If you are the leader of a nation, you are just as impotent in this regard as the bacteria you cannot wash from your hands. If you are a citizen of this planet, you can do nothing to prevent this end from coming.

The only thing we can each do is to come to terms with it and, perchance, with love and understanding of one another as our guide, face this with heads held high and know that we will be coming home. This cataclysm may be the end of this universe, but it is not the end of our existence - not if I'm right. Our immortal souls will continue to exist in the realms that are, now, unseen to us but are, as I've said, always just around a dimensional corner. Finally, when that time comes, we will get a chance to see around that corner. But before we do, we must face this cataclysm together. If our sacred texts are correct, we will all get to see it, even if we have died before it comes, as the resurrection of the dead will occur just prior to that end; so, too, will we get to see all our loved ones again - either before or after the event. Of course, there is the judgement of our earthly deeds and how we act now, before that event, is why we must change how we act towards one another while we still have time to do so.

We must strive to love one another and work with one another. Put our petty bickering behind us and be the best of our species. Individually, that means to be the best 'us' we can be. If we do that, we CAN hold our heads high and face a universal onslaught with the knowledge that we tried our best. But if we continue to fight one another and kill one another for whatever reasons - and there are very few good reasons - then we condemn ourselves and have only ourselves to blame for any discomfort we cause one another and ourselves. We can either act with charity and kindness and act towards one another with love and respect knowing there is little difference between us or we can continue the charade and act out of the ignorance of our animal past and our animal instincts. We, as humans, have instincts that are different from most other animals on this planet and we can see opportunities to help others and be less selfish if we open our eyes and act on those instincts of compassion and brotherhood.

Our science has proven that we are all related - that if we trace our ancestry back far enough, we do have one father and one mother. Although science is not convinced they knew one another. Some sacred scriptures state that there were two who knew each other. Ultimately, though, at this point in time, it doesn't really matter. It is enough to know that, in established fact, we are all family; and, it is with that knowledge that we should warmly extend our hands to help one another where and when we can.

The best advice I can give has already been given before. I can only repeat it. Love your Creator with all your heart, all your soul and all your might; for, in truth, there is only One. Love one another as yourself; for, in truth, there is only One. Do unto others as you would have others do unto you; equally, what is hateful to you, do not do unto others - for in truth, there is only One. Believe wholeheartedly that there is only one entity that exists and you and everything around you are an integral and vital extension of that One. And, most of all don't just believe that way - don't just say it - but act on it. Put it foremost in your mind and bring that spirit of togetherness into everything you do. We are all in this together; there can be no doubting that. Without science and without religion we can still act friendly and warmly towards one another. Let neither of those be fetters to bind you from your own compassion. Religion was never intended to bind compassion but to focus it. Science was never intended to blind us but to open our eyes. So, open your eyes to the world around you and embrace it while you can and try to improve it while you can and endeavour to help those around you do the same to the best of their ability.

Whether you are atheist, theist or agnostic, these acts of compassion and love won't hurt you nor will they upset your sensibilities in any way. Rather, they should appeal to the best of them at that human, gut-instinct level of compassion - that part that makes us human. I appeal to the best part of each of you to do this, whether you are ill or healthy, poor or rich, weak or strong. All these conditions are temporary and will pass. Know that each of us does make a difference - every minute of every day - on everyone and everything around us. This, too, is scientific fact, namely, the Butterfly Effect or Perturbation Theory. So build it into your mind and programme yourselves to help rather than hinder, aide rather than thwart, give assistance rather than walk away. Saying that, there are times when it is best to walk away; but, that is done in order to prevent needless aggression from which neither party would gain. For the people you hinder, thwart or walk away from are your family. So, how could you do that? Would you want them to treat you that way? No, you wouldn't. So, you don't do it.

It's so easy to say, so easy to write and, seemingly, so desperately impossible to organise on a worldwide scale I feel almost ridiculous writing it. But I'm compelled to because I, for one, care. I can't reach each of you and spend hours with each of you like I wish I could. Rather, I can only hope you read these words and let them enter your heart as they have left mine. Who am I to try to convince the world to right itself? Well, just someone

297

who cares enough to try and dares enough to do what I can. It is nothing more than what I believe many of you would do, if you had the time. I believe in you.

Now, I know that may sound ridiculous to many of you. How could I possibly care so much about people I don't know or believe in people I don't know? Well, for all I know, I may meet you tomorrow or I may be your 5th cousin twice removed and neither of us realise it or I may be your 151st cousin 275 times removed; it doesn't matter because I know we're family and I ought to care about my family. The main reason, though, is that I'm an unrepentant idealist, both in the modern sense and the Platonistic sense. Plus, I'm a philosopher whose philosophy demands that I care about humanity to that extent and that philosophy is so heartfelt that I cannot help but act upon it and feel the feelings that living that philosophy demands I feel. Personally, I feel I have no choice but to feel so strongly. I wish everyone could, for a moment, enjoy that feeling or that I could, for a moment, share that feeling with you. If I could, most of the world's problems would resolve themselves in a couple of years.

If you are an atheist and staunch empiricist, then you may consider my theory curious and, whilst possible - even plausible - cannot be empirically proven and so must be relegated to the dustbin or become tomorrow's fish-n-chips wrapper. Note that, as I write these very words, there is one British national newspaper that leads with an article stating that Prof. Stephen Hawking has declared that God is redundant and not required, given the laws of physics, to have a universe spring into existence - yet he can't explain how consciousness arises nor can he explain quantum-entanglement given only 4 dimensions; neither can he explain how dogs can sense earthquakes before they happen or how those 'Laws of Physics' came into existence - rather, he just grants that they 'are'. This strikes me as 'explaining away' the universe rather than explaining it, which is what he supposedly stands against. At the very beginning of this book, I proved his statement to be illogical anyway; nevertheless, I have great respect and admiration for Prof. Hawking and would love to discuss matters with him. Another newspaper leads with an article about the horrible scandal involving fixed international cricket matches. Whilst the real story this (that) day was the on-going flood in Pakistan that was killing hundreds daily and threatening hundreds of thousands. What good-hearted and caring atheist could give a damn about Prof. Hawking's opinion about God (or mine!) or the state of cricket when people are dying and the world is doing little to nothing about it because Pakistan is viewed as a failing State that is nuclear-armed and is purporting to harbour anti-Western terrorists? A little more compassion from Western nations towards these people would show the few who may have sympathies for and with the Taliban that 'The West' is no enemy and that the Taliban are perpetuating a war no one wants or needs. While the West sits on its hands and its food and water, the Taliban mobilise and feed and water as many as they can, thus demonstratively showing the compassion that the West says they don't have. It may well be for political gain, but at least they are saving lives! I ask atheists only to act out of their normal, ethical compassion. I would, of course, love to convince a few on the way, but I feel that if they can be mobilised to act for humanity's sake, then that is almost as sweet. For if I'm right, then their actions of compassion will have saved their souls and God's mercy will prevail even against their disbelief - given enough good, humane acts and benevolent intentions.

If you are agnostic, as I surely once was, then I ask you to consider that my theory explains more than any current theory and, therefore, stands as more likely to be correct even given the fact that it cannot be, at this time (or perhaps ever) be scientifically, empirically proven or disproven. But the bottom line isn't the theory itself but the philosophical implications of it that matter more. Thus the importance of acting as a brother or sister to everyone you meet. It is compassion and the empathy that it brings that is the true bottom line. Maybe you aren't convinced about whether or not God exists but you can hardly doubt the daily struggles of your neighbours and your own. And those struggles could be lightened by sharing the load. This isn't new philosophy but a reaffirmation of what humanity has known for ages. What I've tried to do is to put together the concepts of Plato, Einstein and a few others and, given a string-theory basis, induce the reality that accounts for both the seen and the unseen. Help your family - the larger family of mankind. If my theory is correct, then it is the best course of action you can take; if I'm wrong, then absolutely no harm can come from helping those around you. For one who sits on the fence about God's existence as an agnostic, this is a win-win scenario. If a morsel of false pride keeps you from accepting Pascal's Wager, then, please, don't let a similar scrap keep you from extending your compassion and accepting my wager.

Advice to Individual Countries

United Kingdom

I only begin with the United Kingdom because it is the country of my current residence and I feel obliged to consider and address the place I call home, first. Although I've lost no love for the country of my birth, which I could never lose, the U.K. has inspired me in ways I can barely believe. It's a magical country that, completely unbeknownst to the local populace, is frustratingly close to utopian. As a stranger in a strange land, I was welcomed here and life moved on as if little had really changed. As I've lived here, I've noticed that just a few small changes and the country could be so much more productive than it is now; but, it has to attend to its own population. It needs to know who the people are. Is the United Kingdom ready to accept that challenge and make it good?

It seems to me that the key to kick-starting this country back into full gear is to fully utilise its population. This is not a large country and there is no reason why any person in the UK should ever end up living on a park-bench or in the door of a closed shop. Homelessness and joblessness are solvable and addressable in today's United Kingdom; but, is it brave enough to face the imposition? How can you call yourself a First World Nation, as surely that would imply that it were the best kind of nation; but, can a best kind of nation be 'First' if there are homeless? Don't feel alone UK, America is just as guilty as any, along with many others, of this type of arrogance. But solving it is not out of reach; you simply need to face the problems of homelessness and unemployment and deal with them and surely that's just common decency.

I would like to suggest a few provisions that I believe should be enacted into legislation. These will be socialistic provisions and would be implemented in order to curtail homelessness and unemployment. With zero-percent homelessness and zero-percent unemployment, the nation would be as strong, economically, as it ever possibly could be and then it would be a First World State. Revenues from taxes would be maximised because everyone is working and, therefore, paying taxes; and no monies would be leaving State coffers to support those who are unemployed or trying to give shelter to the homeless. These provisions are not without cost, though, but the results far outweigh the impositions. With regards to those who have disabilities, they will be accommodated as is best possible. Depending on the disability and the mental health of the individual, most can contribute marvellously. I think this is high-lighted by Prof. Stephen Hawking. The fact that, given his condition, he can contribute his thoughts and that those thoughts are invaluable is proof positive that even the most physically disabled can be very useful. Each individual can, in some way, contribute and they exist for a reason. Ideally, everyone should be cultivated as to be most effective. That would be the state of affairs in a First World State.

Firstly, it should be enacted that each Member of Parliament take a homeless person from their constituency into their own home and provide them with shelter. The person will be given the home address of the MP and, therefore, will be able to be registered to vote. The individual will be given a health check and, if that person is found to be addicted to any drugs or have any condition that can be treated, then the person will be provided with a spot in a clinic in order to dry out or to treat any treatable condition they may have. If the individual is disabled in some way, that will naturally be handled in due respect of the overall health of the individual. Once the individual is considered as healthy as they can be, then they will be given support in order to find employment for them. This will be made easier by the provision of the second enactment. Once employment has been found, then the individual needs to have a residence found for them in which to abide. The consideration of job location with respect to housing location should be investigated thoroughly so that the individual, while having sufficient shelter, is not so far from their job that they have no disposable income. In a First World State that is fully online via the Internet, working from home is a very viable scenario, especially for the disabled. In other words, the State, with assistance by the forces enabled by the second enactment will, together with the guardian MP, should endeavour to maximise the individual's disposable income, therefore making the individual most likely to be able to spend that money and keep the cash flow increasing the GDP of the country. There will be, sadly, still those that must be completely taken care of and that can never be fully prevented. In a First World State, anyone within its borders can find food, clothing, shelter and healthcare; the processes of a First World State would never forget anyone.

Once an individual in the care of a guardian MP has found employment and secured their residence, the individual moves into their new home to begin their life again and the MP takes on a new individual and the process repeats. The process could extend down to local Councillors and/or extend into the House of Lords. In fact, I'd suggest each Lord with the right to sit in the House of Lords and irrespective of whether or not they take up that right, they are, nevertheless, responsible for caring for two individuals purely because the average means - the bottom line - is that a Baron has more financial resources than your average MP. It would also make a generous gesture that the members of The Royal Family take on 3 once the age of 18 is attained and the generosity of that family is well attested.

Imagine the press regarding the homeless prospects for which Lord Alan Sugar is responsible! Now that's reality TV... watching Lord Sugar not finding himself a good employee as usual but finding steady employment and housing for real homeless people. This would be just as good for those wanting to exploit the on-going merits of Sir Lord Andrew Lloyd-Weber. Those groups listed above, I should think, won't like the idea of bringing strangers into their homes and providing for them because of the imposition. That is, of course, understandable, but this is not a permanent situation, as there are a limited number of homeless people and, with everyone working together on it, the problem becomes the answer TO the problem. Perhaps it is best, now if I discuss the second provision.

As a second provision, I would like there to be created a department of government that will be responsible for knowing the strengths and weaknesses of the UK population. A real cohesive effort needs to be made to combine Social Welfare, Health, Education, Housing, Employment and Human Resources on a National Level. There needs to be a combined effort at discovering, recording and utilising the talents of our populace, for it is the people that form the basis of our workforce. We need to discover and understand the talents of those who are unemployed, as well as maintain comprehensive knowledge of the jobs available at any given time. All this needs to be coordinated and overseen by the government to ensure the system is working - this, too, will create jobs. To be fair, this secondary provision is so big that it deserves a book in order to explain how the coordination is fully deployed. I raise it here, in short, in order to demonstrate that, with the right information processed more thoroughly and more centrally by people who actually know the individuals involved would be a more effective way of finding employment for our unemployed than the systems in place at the moment.

There is room for a type of *Star Trek* "Counsellor" in today's world, a person who gets to know everybody in a neighbourhood and can be fair and trustworthy and be a direct mouth-to-ear from the constituency to the central government. It's also a job that entails, most likely, the privilege of a certain degree of confidentiality - there must be trust between that

Counsellor and the citizenry - and that's a tall order to fill. I'm not talking about a trained, Government snitch, but a person who can keep confidence and actually cares about the people. This is an ideal solution and the reality may be a group of four people that know different aspect of their lives, say health, education, work-experience and the individual's abilities and aspirations, etc. and that can be further managed by a specialist. The point is that a new network of resources that manage the human resources of the nation needs to be created that can fully maximise the GDP of this nation and make it as strong, economically, as it can be. That is what a First World State would do.

The general process for the second provision is the same as the general process for the helping of a homeless person back into gainful employment: health, employment and housing - in that order. A combination of the NHS, the Jobseekers and Benefits Systems and a better connected UK-wide Estate Agency; thus, perhaps lending the acronym, Health-Occupational-Provision-and-Estate: HOPE. It is hope that this country always says it wants to provide and it has the resources; just at the minute, though, their not tied together properly. That won't take forever to sort out and the payout is maximised GDP, a goal that would make any party who produced it a sure candidate party to continue leading forward. But who will pick it up? Labour because it's Socialist? The Conservatives because it maximises GDP and what could be more Conservative but to retain a position of the conservation of total employment throughout the nation? Do the Liberal-Democrats pick it up because, well, it's a nice thing to do? It seems to tick everybody's box in some respect. Maybe that's because it is the right thing to do and the right thing to do crosses over all the political spectra in a democracy because, in a true democracy, the government only runs because the people ask it to do so and the people would not act to harm themselves.

You cannot allow your fellow to become or remain to be fallow. Surely, any decent, healthy nation and First World State has, by virtue of being its guarantor, a responsibility to provide the basic dignity to each member of its population not to have to sleep in the streets. Of course, an indigent may not be able to provide proof of citizenship; then, dignity itself deigns that any fellow human has the right to protection from a decent, First World State and humanity is the test rather than citizenship. You may fear this will increase a flux of people from other nations but this is already the case because Britain has always stood for these values in the eyes of the oppressed. If Britain has ever stood for anything, it has stood for dignity. Then provide it! You have the means. It might be easier than you think, as many of the first people provided for will help bring in and help others. It's just a matter of getting the ball rolling and the measures, above, creates that ball and pushes it down the hill. If you want to call yourself a First World State, then be one and let the human dignity for which Britain stands be its greatest test and greatest victory.

China

I think it's fair to say that most people are afraid of you, which, naturally, is both good and bad. It is certainly good from your standpoint as it gives you great confidence that you will continue to prosper. It is bad if you cannot handle the responsibilities of what you could attain. It is, again, fair to say that, at the rate at which you are buying American Treasury Notes and Bills, you will soon control the American treasury. You may think this gives you certain powers to establish American policies by withholding money from policies deemed unworthy and supplying money for those deemed worthy. This, though, is a huge responsibility. Can you handle it? Can you act wisely? For if you do not, you will find yourself overstretched and will crumble. Alternatively, the worst outcome is war. People may fear you but you are not invincible.

There is almost no other country in the world that can mobilise its people with the efficiency of China. It is incredibly admirable that so many differing people have united for common goals and achieved such success. You have taken the basic concepts of communism and learned how to and have successfully turned the capitalist environment around you into a playground in which you accrete wealth through their system while maintaining communist applications of the wealth. Over the last 50 years, no other country has been so successful. What more could you want? I suspect the answer to that question is, how much can we have? You can have China. But what is China?

I think you see where I'm headed. Tibet. Is it really China? What's wrong with a China that does healthy trade with an independent Tibet? Both countries could benefit from Tibetan independence. The resistance will never stop and the cost of the administration of an hostile area is not worth the price. It is more prudent to let Tibet be Tibet. The same, of course, is applicable to Taiwan for exactly the same reasons. Plus (and no offence to

Tibet), decent trade with Taiwan couldn't possibly harm China in any way, shape or form. Again, it's a win-win situation. It's time to let these things go and focus on China. The world is a different place, now. Besides, you've regained Hong Kong; it is not wise to be greedy. Continuing to hold onto Tibet and trying to regain Taiwan are both lost causes. It is in China's best interests to let them go.

If, in time, you gain control of America's treasury, you must be very careful how you deal with the responsibilities that go with handling America. The American people are not Chinese. They will not tow the line and will make keeping the peace in Tibet (and Korea and Southeast Asia) look easy. As you know, most of them are armed and, in general, the Americans are an unruly lot. I should know; I am one. When the Americans fight, they tend to, if possible, fight dirty and that should be your greatest fear. Remember that they won their independence by using guerrilla tactics against soldiers who lined up in ranks and files. They handed out blankets laden with small pox to the Native Americans in order to infect them and reduce their numbers - clear genocide. They are the only country who has ever used nuclear weapons and they did so knowing there could be no like response, as they were the only possessors of such weapons.

If you handle them in such a way as to engender their wrath, they will turn nasty. I reckon that most American and Chinese strategists have realised that, in order to engage China in a battle, America, most likely, would have to attack them from the coast. This China can easily defend against, as they can always retreat into the hills and stave off any troops who venture too far inland. It would be 10 times worse than Afghanistan. So, the Chinese sit back and laugh and think they are invincible. But it is not the case. "If the van is strong, attack from the rear". The way to beat China is to attack from the rear, as both sides already know. But to attack from the rear means not fighting in a conventional way, it means fighting nasty and that is what America has done, can do and can do well.

Naturally, a nuclear attack is useless as it would mean total devastation of both countries in a mutually assured destruction scenario. No, the answer, as both countries probably already know, is a biological attack. You have to drive the people out of the mountains and towards the coast while, at the same time, ensuring that returning to the mountains is impossible. This can only be done by the use of biological weapons and the poisoning of China's rivers with some agent like botulin toxin. This would force the population towards the coast packing a starving and thirsting population together into barely liveable conditions and THEN an invading navy could pummel the coast conventionally, attacking the starving and thirsting sitting ducks. This is ruthless and heinous and I do not recommend this in any way; I state it because it's the truth and China must acknowledge that they are not invincible. Only then can they recognise the situation they are encroaching upon.

The responsibility of owning America's treasury is not as easy as you might think and I hope you now realise how ugly things could get if you get it wrong. I would, of course, denounce such a criminal attack on any people, as this type of warfare is horrendous beyond comparison; but, that is exactly the way America can fight if it feels it needs to do so. So, be warned. It would be one of the most - if not the most - heinous acts of war ever perpetrated; but, as I said, this is exactly how America fights when it wants to: it gets nasty when it knows it can. Japan, I'm sure, would agree, as would the Native Americans who were their first biological targets.

I would hate to lose China. Its land and its people are irreplaceable and they are an enormous gift to the world. I want to see a prosperous China, which is why I point out the potential scenario above. I don't want that worst-case scenario to happen. If China wants to own the U.S. treasury, then they must act responsibly once it is in their possession and not let America suffer in any way. China could improve many things in America but it cannot make it like China politically. Americans are a different people and they're not afraid to be vicious, especially if they feel their backs are up against the wall. China could help America and a closer relationship between the two could be a wonderful bonus for the planet, as better relations ARE better relations; but, it will be down to China, rather than America, to determine how any American money is spent if China owns the American treasury. If that is not handled with the utmost delicacy with respect to the American people, the result could be beyond tragic and I don't want that to happen. Please be careful.

Note to the Chinese Government: I realise that you will receive and understand the criticisms and suggestions above even given only a publication of this book in English. I would like you, the Chinese Government, to understand that I am willing to create an abridged version of this book that contains NO criticism of the Chinese Government whatsoever; so that, hopefully, you will allow that abridged version to be made available to the Chinese people. It is this author's sincerest wish to cause no internal strife to the great nation that is China. Rather, I only wish to help. The last thing that China needs is civil strife brought about by the criticisms I put forward here and, in truth, the world would not benefit from internal strife in China either. The above criticisms and suggestions (and other comments made in other sections) are directed only towards the Chinese Government not its people and, as stated, I know that this book and its criticisms - if only ever printed in English - would reach its intended audience, which is the Chinese Government; therefore, I am perfectly willing to make the concession of creating a version (at a later date, so there is time for discussion regarding this if you wish) translated into Chinese that does not contain any criticism against the Chinese Government and that version should be viewed by the Chinese Government as being suitable for the general Chinese population.

I firmly believe that the vast majority of this book is of no threat whatsoever to the stability and integrity of Chinese politics and, because of that, I would like the Chinese people to be able to read it and benefit from the ideas contained herein. Obviously, China is an important player in this world's political arena and an enormous resource that could benefit the entire world. Notwithstanding, I do not desire to be considered a political enemy of China and no one in their right mind would. I hold no enmity towards China at all; as I said, I only want to help if I can. Besides, in my opinion, your country produces the best cuisine I've ever enjoyed! Please consider this as a potential compromise that would prevent this work from being censored in your great nation. Thank you.

United States of America

Firstly, I believe that America should adopt the same laws that I have proposed for the United Kingdom so that the American unemployed and homeless are brought back into the system. Of course, there would have to be some minor changes and some major changes in order to allow that to happen. To begin with, America needs socialised healthcare. Having lived in both the U.S. and the U.K., I can see the enormous benefits that a National Healthcare System can provide. I believe that every government has the duty to provide for their people and one aspect of that is maintaining the health of its population. The scaremongering that socialised healthcare is somehow politically dangerous or is just one step away from downright communism is not only ridiculous, it is an outright lie. Most of the people who say such things couldn't give you a proper definition of communism, anyway; they don't really know what it is. But National Healthcare is about the people caring about the people. Shouldn't Americans care about Americans? Who else will if they don't... the Chinese?

And don't think I haven't seen the hypocrisy - in the people and the government. You are 'oh, so proud' of the Statue of Liberty but seem to have forgotten what dear Emma (Lazarus) wrote on it: "Give me your tired, your poor, your huddled masses yearning to breathe free." Yet you worry about immigration? America invites it. America was, is and always should be a place to which to run. Immigration is what made the country strong in the first place. Different people doing different things providing the various services a nation needs. Sometimes you went so far as to wage war on indigenous people to make way for the growing population that was growing largely due to immigration. Hopefully, you've learned that was wrong and will now own up to the responsibilities you, like Britain above, have to your population.

There's an internal threat to America's stability, too. This is the legacy of J. Edgar Hoover. Whilst, to many people, this may sound like just another conspiracy theory, it is an historical truth. The country is run from behind the scenes. Every President since Truman has not really been fully in control of the country. Truman, himself, saw it dissolve before his very eyes. Why do you think the papers had already declared him the loser in the re-election? The press had been told Truman would lose by those who had conspired to try to ensure it. Thank God, though, there were enough voters that made that impossible. Hoover conspired with Nixon during the Eisenhower days and the both of them saw the opposing Democratic Party as an enemy rather than a political opponent with differing policies. Ever since then, the office of the President of the United States has been less of the 'Face of Government' and more the 'Façade of Government' - excepting, of course, the Nixon/Ford administrations. Both Kennedy's lost their lives due to the fact that the FBI was complicit with organised crime and taking their cut. Bobby knew it and wanted to stop it but they fired a shot across his bow with the JFK assassination and debacle. LBJ was horrified when he had to take over, as he knew the truth and knew there was a gun always pointing at his head. When it looked as though Bobby Kennedy would become President, he had to go. That paved the way for Nixon to take the reins personally. J. Edgar couldn't have been happier. Nixon, then, tried to get inside information on the plans of the Democrats (Watergate) with full FBI backing (they were the ones who planted the bugs!) and won re-election; but he got caught. Agnew ran away before the metaphorical shit hit the fan and Ford was brought in. Ford was told the 'truth' about the whole plutocratic oligarchy that was running the nation from behind the scenes and it was decided that the CIA was better placed internationally to be the front to the New World Order.

So George Herbert Walker Bush, already an insider, was appointed as head of the CIA by President Ford and the power

313

behind the power was handed over to him. If you noticed, this is when the movement changed direction. No longer were there the internal witch hunts that McCarthyism brought - the type of enemies that the FBI could ferret out - but the new enemy were foreign outsiders hiding in foreign lands (the realm and remit of the CIA): terrorists. Carter never stood a chance. 444 days of not being able to retrieve hostages from Iran finished off the Presidency of Carter, one of the most honourable men to have ever served in the office. Once Reagan was elected, Iran was informed that they would be targeted if the hostages were not returned and, lo and behold, they returned just in time to see Reagan take office. The Americans were happy; but they'd been duped. Having a real actor, Reagan, as a sock-puppet was a godsend to the system-behind-the-system and, after Reagan's re-election, it became a platform to launch G.H.W. Bush into the White House itself...and so it was. But his direct reign didn't last long. It was long enough, though, to point out Iraq as a 'rogue nation' that 'needed' to be handled.

Clinton was bright enough not to fight the system behind the scenes and was given extreme leniency as a reward. Then Bush Jr. came in and, again, plays muppet, while there are voting scandals during both of his elections. Had Florida (in the first election) or Ohio (in the second election) gone the other way, Bush wouldn't have won. But the winners write history and, most likely, vote early and often. Nevertheless, he tried his best to 'smoke 'em out', those evil, terrorists. We moved from J. Edgar Hoover fighting for and against the likes of Al Capone and, now, America does the same with Al Qaeda. Both Saddam Hussein and Osama bin Laden were groomed by the CIA and aided and abetted when they were deemed useful; but, when they became expendable, they became the face of the enemy. The NSA (National Security Agency), which is supposed to watch both the FBI and the CIA to ensure they are acting in America's best interest, is culpable by doing nothing to prevent the illegal wars and atrocities

314

perpetrated by those behind the scenes. In fact, the NSA is riddled with 'Yes men' that actually make it easier for these operatives to take these actions.

Now, we have Obama as President. But did you notice how his face fell just a couple of weeks into office? It was like watching LBJ all over again. Someone informed the President that, in fact, he wasn't really the leader of America and that he would have to tow the line or face enmity the likes of which he had never imagined. Every former President has the right to daily updates from the CIA, but only George H. W. Bush claims that right. He is still calling the shots and is, perhaps, well placed to quietly do so. But the next President, after Obama, will be the one to fear the most, as, most likely, that President will be a Republican and an insider because the Neo-Cons will destroy anything good that Obama does and will drag his name through the mud such that America will want change; it's exactly what happened to Carter and Obama is just as much of an honourable man as was President Carter. But the Republicans are already dragging his name through the gutter and doing it well.

Since Reagan, the Presidents of America have followed a trend not dissimilar to a repetition of the trends set by the Caesar dynasty of Ancient Rome. Both Julius Caesar and Ronald Reagan were given the appellation of 'The Great Communicator'. George H. W. Bush followed closely on the coattails of Reagan in much the same way that Augustus Caesar followed Julius and he also consolidated power in much the same way. Then came Clinton, like Tiberius, he must have thought he was God and could get away with anything (Monica Lewinsky) and did. Then we were given Caligula: Bush Jr. A short, foot-stamping loudmouth 'little boot' who thought the Presidency was, more or less, a joke because daddy's really running the show. Now, we have Claudius in Obama. Like Claudius but for different reasons, Barack Obama, being black, was an unlikely candidate for the top job, yet his policies make sense and, like Claudius, he is of an incredibly kind nature. So he's sure not to last long. After that, there comes Nero. That's what is facing America after Obama: the return of a Nero-like figure.

Nero or whoever is President after Obama, is going to ruin America so be careful for whom you vote. It's a case of taking up The Who's warning and *Don't Be Fooled Again*. "Meet the new boss... same as the old boss." The Tea Party would love to be your Nero. Even Sarah Palin would love to be your Mrs. Nero. After all, she barely fiddled with Alaska before she was almost a heartbeat away from the White House; she'd relish the chance to fiddle with the entire country. Don't get me wrong, though, I believe Sarah Palin actually means well but she does not have the intellectual wherewithal to do anything other than to take the advice of her advisors and THEY are the ones who would set policies; not her. After Nero, the empire began its fall. So, too, no great State lasts forever and America will be plunged into riotous anarchy as the plutocratic-oligarchic insiders safely watch from their gilded bunkers. It wouldn't surprise me to hear the President, at that point, say to the unemployed and starving masses, "Let them eat Twinkies!"

The Tea Party and the Neo-Cons are, albeit dressed very nicely, simply extremist fronts. These are the 'America is for Americans!' people. OK, give it back to the Native Americans. As The Temptations told us in *Ball of Confusion*, "... and the only safe place to live is on an Indian Reservation." Well, I doubt very seriously if the 'government' would do that. Their problem is with the definition of the word 'American'. What does it mean to be American? Certainly, you are American if you were born in America or are a naturalised citizen as provided by the Constitution. But those that come to America, wanting to be American, are already an American in their heart. Most immigrants really would rather be American considering their current situations. That's why they're prepared to leave their homeland and, like generations of others before them, they embark on an adventure to America, holding in their heart—oftentimes greater than so-called natural citizens—

the American Dream, the hope to eke out a better future for themselves and, in some cases, their families, as well.

If the government or any party wishes to prevent immigration to America let them remember that it was the power of immigrants that put railways across the nation. Immigration is the soul of America. America's strength, like humanity's itself, is in its differences. Difference allows and demands change and change is the means by which progress is established. Without change there is nothing but stagnation. But certain groups actually want this.

You may have noticed that I used a couple of musical quotes, above, to make certain points. This is because music often reflects the events of the times. I clearly remember crying once I'd learned what the Crosby, Stills, Nash and Young song *Ohio* was about: the massacre at Kent State. The powers that were in charge then are more hidden but they are still there. The rap music of today seems to be generated by a class of people largely forgotten by the system and they try to govern themselves in gangland style. It isn't pretty. Plus, it asks questions about a system that DOES forget its people and does little to prevent the establishment and maintenance of gangs. Why? Well, there's a lot of profit that can be drawn off being on the inside. Al Capone and the old gangs were a training ground for the modern equivalents like the Cripps and the Bloods. These groups are powerful and are linked in many ways to foreign gangs like the Yakuza and Tong. There are streets in America where these gangs control acres of cities and the government doesn't stop it. Why? Because it's in on it.

Very subtly, this behind-the-scenes government has infiltrated almost everywhere and certain people take their cut and all is hushed. As long as the rich get richer and the poor die off and most people are too stupid to realise what's going on around them, then all is fine. Wake up, America! The last time I was in the States, I was shocked watching the news. Not because of terrible events but because of the subtle brain-washing. The Press is no longer free in America. Not that you get much better in the UK. In the UK, you get huge headlines about what sports star slept with how many women that was not his wife and small headlines about Syrians hunting for food. The BBC gives a prim and proper spiel of BS as opposed to the cutesy, 'Katie Couric' (apologies, Katie, I know you're just doing your job and you can't help being cute) style in America, but it is BS just the same.

American citizens are being fed a mixture of lies and dressed-up truths. How can you tell? A simple test might be whether or not you take a dislike to Muslims (or Mexicans or Afro-Americans or Native American Indians or any other people) when, in fact, you've never met one. I suspect many people are hanging their heads in shame, now - or should! If you do take such an automatic dislike because of what you've read or heard, then you've bought into a lie that Muslims (or any of these 'other people') are all like Osama bin Laden. Osama stopped being a Muslim the first time he ordered an attack that killed innocent people and that was the end of his Muslim days; once he had gone that far his challenge became "how many innocent lives can you take?" Anyone who would think like that is no Muslim by definition. If you take an automatic dislike to any people different from yourself you've fallen into a trap and you're listening to rubbish. America is about difference and the strength it offers. Don't EVER forget that. The enemy is within and the question is, "Will the American people be able to free themselves from that oppression?"

The truth is: People are people. Most people just want to work and live a quiet life no matter what their ethnic or religious background is. America was founded on the basis that humanity was what linked us together and that dignity is to be shown to every human. But in the streets of Harlem and Watts and East L.A. there is no dignity. Police say, "They can't fix the problem." This may be true, as the government doesn't care. Capitalism doesn't care. Socialism, though, does. If America is to wrest itself free from the fetters of occult government, it has a very hard fight on its hands. It may lead to a second civil war. It would be better to take a more Gandhi-like approach or a Martin Luther King approach. But we know what happened to them. You have to be brave and isn't America the 'home of the brave'?

Big Business is a huge part of the problem. Rampant capitalism in the hands of those already morally challenged is deadly and ruthless. Pharmaceutical companies and oil companies have their hands in government. You know that oil is the family business of the Bush's; why can't you open your eyes and see what's happening? Oh yeah, I forgot, your free press is numbing your minds. The War on Drugs was, is and always will be a hugely dramatic failure to prevent illegal drugs from entering America because the government is fully aware of the flow. The DEA only does what it's told to do and it isn't told much nor is it unmonitored. Ruthless capitalism is corroding the heart of America. Money means more than life itself. Well, excuse me, but I don't remember the Declaration of Independence stating that we had an inalienable right to money and nowhere did it state that the pursuit of happiness overrides the rights to life and liberty.

Christ, what a polemic! But is that all there is? Is there hope for America yet? Of course, there always has been. America is all about how to work together. So, who are the hidden enemies of America? They are those, from within, that want to separate these people from those. America is about unity; that's why it's called The UNITED States. Not just united by being next to each other or speaking the same language but by the common factor that they are each different and, so, can lay no claim to superiority. I was taught that Americans were all equal, that is to say, that they were equally individual. We know from the philosophical implications of the physics of Special Relativity that each of us is vital; so, its time to work together and to do so honestly.

But how do you get rid of the corruption and, at the same time, preventing an updated version of the Fall of the Roman

Empire or a combination of that type of fall with that which happened to Athens ending in the period of the Thirty Tyrants? I'm unsure. In fact, whilst I firmly believe that such dissolution can be prevented, I'm not sure of the methodology. The problems that face America differ from those of ancient Rome and Athens and not by a small amount. Perhaps it is to refuse the circuses aspect of the bread and circuses that will be offered to propitiate the American people. The circuses are already being proffered, by the way: The X-Factor, America's Got Talent, etc., and, of course, the same is true of Britain. All these programmes are devised to entice, entertain and pacify a people slowly starving. It's worse in Britain where they demand that you pay for the privilege of watching TV. Perhaps the key is to demand more bread and damn the circuses! Demand that government provide what we all know it can.

There would be no homeless or starving in a first rate or 'First World' country. So, if America wishes to lay claim to such a title, then it must needs provide for its people - all of them. A first rate country would care for its people and not forget any of them. There is little doubt that protectionist policies will have to be drafted. You need to protect yourself from China and one way of doing that would be to only sell Treasury Notes and Bills to U.S. citizens. Granted, this is very protectionist and prohibits other nations from investing in America; but this policy needn't last forever, rather, it only needs to last as long as the self-defence it affords is required. Employ similar measures to those I've proposed for Britain in order to remove homelessness and unemployment and this, by its very nature, will require a system of socialised healthcare as I've stated above.

For America to retain - or, indeed, attain - its claim to be a first rate or 'First World' nation, it needs to put its money where its mouth is and rid the streets of homeless and unemployed citizens. It must keep itself open to immigration, as an influx of people that are put to work will only increase the national coffers via taxation and this will, in turn, create a stronger America; this is basic economics. But Americans must rid themselves of their cancerous racist beliefs. There are, in fact, many 'isms' that need to be expunged: racism, genderism, ageism, sexualityism and any state of mind where some petty difference is blown out of proportion and a person or persons are made to suffer simply for being different. Our differences are our strength! The American government must protect its citizens or, by its own laws, it must be supplanted by a new form of government that CAN provide for its people. If you disbelieve that last sentence, reread the Declaration of Independence: "...that whenever any form of government becomes destructive of these ends, it is the right of the people to alter or to abolish it and to institute new government." If the government, at that point, claims that the Declaration of Independence is not legally binding, then we are still colonies of Great Britain and you, the people, have caught them in a major lie; not to mention that it would be an outright extortion of liberty if the government turns a convenient blind eye to parts of that Declaration.

The worst way forward is, of course, a second civil war. This would cause a huge amount of death that might be avoided but, if worse comes to worst then the people may be left with little choice. At that point, the Second Amendment becomes the basis for strength to the people because it provides them the right to bear arms. America is a difficult country to invade because of the right to bear arms and it is that right that provides the American citizens with the means to threaten civil war. This would not be a war of secession as the first civil war was but a war of revolution against a tyrannical form of government. It would, in that case, be more like the Revolutionary War (the War of Independence to British readers) but, in this case, the government that is being 'tried' is your own. If your government lets you down, your own laws provide you with the right to overthrow that form of government as I have outlined above; but do not use this tactic unless it is absolutely necessary. A tyrannical government that has nuclear weapons may nuke their own people and this would be disastrous - not only for America but for the entire world. Civil war is a last resort, only to be used when all peaceful forms of revolt have proven unsuccessful.

I fear for the future of America, as I see it well down the path of self-destruction. As a citizen, I have a vested interest in the continuity of a decent form of government and know that a strong America makes for a safer world. But the current ebb and flow of policies have weakened America and, I believe, most Americans are completely unaware of this weakening. The young are no longer as well educated as they were 30 years ago. The standard of education has, like in Britain, slipped and made for a populace that is, largely, ignorant of the important events that take place around it and within it. Gangs take over the streets and the police and government are corrupt. This must be turned around or the country will suffer far more than it can imagine.

More and more people are turning to drugs and alcohol than ever before. Prohibition will never work, though. What will work is to abolish those things that make people turn to drugs and alcohol. People use these to escape from reality for a while. If the government provided an environment from which the people did not wish to escape, half the problem would be solved. If the government provided a decent education about what these substances can do to a person and how they slowly kill or maim, the other half of the problem would be solved. But is the government working towards these ends? I don't see it and, if you don't see it either, then 'We the People' must force the government to provide for the needs of its people. The alternative is that America - quickly or slowly, but, no doubt, at some rate - will disappear off the map and that isn't necessary. It could end up being a wholly-owned subsidiary of China Incorporated.

I love Chinese food, but I love America more. It will always be the nation of my birth and, as a citizen, I feel I owe a debt to it by providing what advice I can in order to bring it forward to a greater glory than it has ever known and it is only because of that sense of debt that I write here what I write. I don't want a second civil war nor do I want homelessness and unemployment to run as rife as it has. These things can be solved, although it may take some hard work, it is well within the abilities of the American people and the American government to solve these issues. As with any nation, I offer my services as an advisor to help in whatever way I can; but, as an American citizen, I feel I owe more to the country that raised me than I do to others. So, Mr. or Madame President, if you want my advice, please give me a call and we can begin to solve these problems - or any others which you feel you could share with me - together.

Iran

Iran, you're a tricky country with which to deal and perhaps even more tricky to advise. I want you to know that I have no ill feelings about you whatsoever. The Persian Empire was, without doubt, the best empire the ancient world ever saw and the rest of the world should remember that. It was the most fair and most just empire and the people have not changed greatly since then. In fact, the only real change has been the introduction of Islam.

America, though, fears you as they believe you to be a 'rogue' nation; that is to say they believe that you are unpredictable and, therefore, must be handled with extreme care if not extreme prejudice. I don't believe you are a rogue nation but a nation that wants to be viewed on an equal footing with all other nations - and so you should be. However, you do seem to have a problem with Israel and that is your greatest stumbling block. I think you should accept as fact that Israel is here to stay and will be, for the foreseeable future, a nation that, like your own, must be dealt with on an equal footing. This seems, though, to be a difficult decision for Iran to make. Surely, you can grant to Israel that which you yourself want to be granted by them: recognition as an equal and independent state.

I believe that Iran can make a huge contribution to the world, as Persia has many natural resources that they can use to trade with other nations and, therefore, Iran stands to benefit from a peaceful environment in which this trade can flourish. Because of my suggestion to Islam that Islamic nations should return to a Caliphate, the Caliphate would need to include Iran. This, in turn, requires that Shiite and Sunni Muslims work together in harmony. This can only happen with the full cooperation of the Iranian people and, of course, the Iranian government. While that is easy to write, I wonder how easy it is to implement.

The differences between the Shiite view and the Sunni view are many but they do not exceed the similarities and that must form the basis for moving forward. Also, of course, Israel must be protected and both Shiite and Sunni aspects of the Caliphate would have to ensure that. The question is, though, can Iran actually do that? At the moment, there seems to be an underlying mistrust and distrust of Israel and that is not without basis due to the previous unilateral decisions of Israel to strike Iranian targets. Don't worry; Israel is not a threat to Iran. I say that because Israel is surrounded by Islamic nations and must learn to live with their neighbours in peace.

At the time of this writing, Mr Ahmadinejad is the current President of Iran and I would like to make a few statements directly to him. I understand that you are a firm believer in the immanent return of the Hidden Imam and I believe that you are completely correct to hold that belief. The signs of the Hidden Imam's return are, literally, written in the stars and, I believe that ancient Persian astrologers would agree if they were here to see what astrological formations have occurred recently and those that will in the near future. In my opinion, I see his return coming from either Qum or from near the region of old Persepolis. However, there are issues that must be dealt with prior to his return in order to make it safe.

Iran wishes to proceed with its technological progress towards nuclear energy and with that comes the likelihood of gaining the technology to produce nuclear weapons in the process. This is what scares America the most. America is afraid that you will gain nuclear weapons and use them against Israel. This would be disastrous for Iran if it acted in such a manner as it would be the end of Iran simply because it would incur the wrath of America, Israel, Russia, France and the United Kingdom - all of which are nations with which Iran could benefit from having GOOD relationships.

Firstly, if it is true that Iran is only interested in having nuclear power, then why not ask for American help? Ask America and some of these other nation to help you towards gaining nuclear power by inviting them into your country to aid you with the technology. This would allay their fears and it would strengthen the bonds between your nation and others and you would regain their trust. If, however, you are seeking to gain nuclear weapons as well, then you have a problem and you will never gain their trust and may be forced into isolation. This would not help Iran or the world in any way. But let me remind you that if Iran enters into a coalition with the other Islamic nations and forms a Caliphate as I have described above, the Caliphate would have nuclear weapons because Pakistan has them. So, you can gain that ability simply by agreeing to enter the Caliphate.

With respect to Israel, you have your reasons to distrust them but I do not believe that a cold war between Iran and Israel will serve either nation or the world and I believe that, in truth, Iran wants to exist in a peaceful world. So, the anti-Israeli attitude must be put aside. If Iran dropped its threatening language then Israel could and would be pacified and this would lead to peace in the area and I believe that Iran would and should prefer peace to war - especially if the cold war turned hot. The years of wars between Iran and Iraq should have taught Iran that peace is better than war and that Iran prospers better when it is at peace with its neighbours than when it is not; therefore, a peaceful approach with Israel would be in Iran's best interests and in the best interests of the entire world. In addition, I believe that better relations between Iran and Israel as well as America would boost tourism to Iran which would also boost Iran's income.

Iran has countless ancient sites and ruins as well as incredibly beautiful natural scenery that many people would love to come and visit and see; but people are put off travelling to Iran because of the political tension. Remove the tension and invite the world to visit your nation. Personally, I would love to visit Iran, but for the very reason I've just stated, I simply wouldn't go at this point in time, as I don't believe that my American passport would be welcomed. Is that true? I'm unsure enough to stay away for the time being but I would love that to change.

There are several specific places I'd love to visit as well as take in the scenery that can be seen no where else in the world. Where else in the world but Iran can you find the Tower of Silence but at Yazd or the ruins of Pasargade and Persepolis or the stark beauty of the Salt Desert? Where else can one seek the shrine of such an important soul as Imam Reza for whom, I personally, have so much respect? Wouldn't it be fabulous to hold the Winter Olympics at Tehran so that the skiing events could be held at Dizin in the Elburz Mountains? Nowhere else can one explore the southern coastline of the Caspian Sea or see the amazing splendour of the views offered by the Zagros Mountains not to mention their historical importance as a barrier and natural border.

These are only a small selection of the vast natural, sacred and historical treasures that lie within the borders of Iran and there are countless others. I believe that by being more open with other nations and peoples that Iran could only benefit from the increase of trade and interest that tourism provides. Iran is crucial to the world and a precious commodity. Please reconsider how you form and maintain international relations such that Iran and the entire world can benefit; please, for your sake, for the sake of humanity and for the planet at large.

Pakistan

Well, where do I begin? Pakistan, it is understating the obvious to say, 'You have problems.' There is corruption at every level of government. In fact, it might be true to state that the least corrupt form of government at this point in time is the Taliban; unfortunately, they are the most dangerous. But what makes the Taliban both less corrupt and more dangerous? They know what they are doing; they are acting with purpose and with goals in mind. The official Pakistani Government lacks purpose and general strategy and every politician has their own goal, namely, to line their pockets with as much gold as they can before being discovered. This plays right into the hands of the Taliban, as they are more than happy and perfectly capable to line those pockets. If this continues, Pakistan will be a failed State and many already consider it to have failed. So, what are you, the official Pakistani Government going to do about it, eh? You need advice, so listen and read.

No nuclear power has ever come so close to becoming an absolute failure at self governing as Pakistan is now. The main problem is that you lack purpose. You seem to have forgotten that for which Pakistan was created: unity. You are a people divided against yourselves and this is why the Taliban can play you so easily. The Kashmiris don't work together with the Baluchis, the Punjabis don't work together with the Sindhis; in fact, none of these individual ethnic groups want to work with any others. I see the very same problem occur in the Pakistani immigrants in the United Kingdom. Those of Kashmiri descent won't shop in shops run by Baluchis and the Punjabis won't let a Sindhi cut their hair. This is ridiculous and childish. You are one people now - both in Pakistan and the U.K. - so act like it! Grow up!

Pakistan was created to provide a place where Muslims can live and work together irrespective of their ethnic backgrounds.

Islam doesn't divide people based on ethnicity so why don't you practice it? Because Pakistanis are Muslim they should, by that very nature, embrace one another fully and disregard the petty differences over which none of you have any control - like ethnicity. Perhaps you should pass a law that prevents the use of ethnic languages and only Urdu is to be spoken in Pakistan. No more will people be allowed to speak Baluchi or Sindhi or Saraiki or Punjabi or Pashtun or Kashmiri. In truth, I think that's going a little too far and it would be impossible to enforce. But you need something to bring back the unity that was the impetus for forming Pakistan in the first place.

You need a common purpose and common goals. You need something for which everyone can work together that can restore that sense of unity. Muhammad Ali Jinnah would not recognise modern Pakistan and would be deeply ashamed of how the Pakistani people and the Pakistani government have completely trashed his concepts. Remember he was President of the Muslim League for several years and deeply understood the concept of Muslim Brotherhood. Now, you have almost thrown it all away. And it is the Taliban who will pick up the pieces if you are not careful. They are already well on their way to doing just that because they understand Muslim unity. Their problem is that they don't understand that Christians and Jews are protected people and Shiite Muslims are brothers of the Faith and that they cannot fight Christians, Jews or Shiite Muslims without also becoming Muslim apostates. They also are willing to snatch 7 year-old girls and strap bombs to them and send them into a market full of innocent people; this kind of action is the most horrific, barbaric, inhumane and un-Islamic that anyone could ever perpetrate. This is why I say they are dangerous, as they simply do not understand what is Islamic and what is not. After all, they are the students (Taliban) not the teachers (Mullahs) and they did not get the best of an Islamic education, as the Mullahs themselves didn't teach proper Islamic truths to them.

If the official Pakistani Government does not eradicate the corruption that is so rife throughout it, the government will fail and the nuclear weapons they possess could easily fall into the hands of the Taliban. This is one thing that must never be allowed to happen; for if it did, they would use them without hesitation and, most likely, the first target would be India. This would begin World War III. It would also bring about the complete destruction of Pakistan because India has far more nuclear weapons than Pakistan. This would be a disaster for the entire world not just southern Asia. Also, the missiles that Pakistan has do not reach very far. If the Taliban were so stupid as to launch one towards Israel, it would fall short and could, potentially, land in Afghanistan, Iran, Turkmenistan or Iraq. Thus, they would be landing on Muslim soil and they would be murdering fellow Muslims - not that they don't already. God help them if they can't tell the difference between West and North and one 'accidentally' fell in Russia. But I don't rate the Taliban intelligence and wouldn't put it past them to make such an ill-fated attempt. This is why the official Pakistani Government must get its act together now and eliminate the internal corruption and prove to its own people, then the world, that it can be responsible and act in their best interests.

Pakistan must promote the unity of its people and it must act honourably. Otherwise, it is simply not an Islamic Republic. Also, if you have read above, you know that I am in favour of a completely united Islamic State by returning to a Caliphate and Pakistan is, arguably, the most important nation within that Caliphate because of its nuclear capabilities. It is the lynchpin that could be the insurance upon which the entire Islamic world relies in order to be taken seriously as a major world player. Pakistan must rid itself of corruption for its own good and the welfare of all Islamic people. It has the chance to set a great standard for honour - that is what Jinnah wanted - but it has squandered that chance ever since his death. It will be hard, of course, but it is far from impossible and this internal corruption does Pakistan no favours internally or with regard to foreign relations. Purge the corruption but do it legally not by means of assassination and more corruption. Bring the corrupt leaders to task, try them fairly and, if one is found guilty, imprison them. Do the right thing and protect your people - all of them.

There is only one Pakistani people. The petty ethnic differences are negligible. Honour those people and forsake the extra gold in your pocket, for it will not buy you into Heaven. It is a great challenge but the reward is what Jinnah wanted: unity. Plus, it has the added advantage of taking away the means through which the Taliban could destroy you. As 'Pak' means 'pure' in Urdu, you must purify your nation. Only then can you truly be called Pakistan - the Land of the Pure.

When Science Meets Religion

Exegesis on the 'Niche Verse / Light Verse' from the Qur'an

Originally, this was a part of the first draft of the conclusion; however it just didn't seem to work there and, as it happened, I had to completely re-work the 'Conclusion' section, above, such that that section was more of a recap and re-assertion of certain aspects of topics from the chapters preceding it. Yet there was a great message in this bit that I couldn't bear to lose; so, I created a better area for it, as it is, once given the book's premiss, a stand-alone derivation from it. This verse of the Qur'an is one of the most mystical and, prior to now, little understood verses. I hope that the Muslim faithful will find my exegesis to their liking. Here it is.

Science has problems with the answer to the question, 'Where did all the anti-matter go?' My concept is that, shortly after the Big Bang, the anti-matter moved along the face of area in which 'that which is' exists. This outer 'face of God', if you will, is set up against the surrounding nothingness (that 'absolute nothing' about which I spoke), which acts as something more solid than any solid, as it cannot permit the existence of anything. The 'something that exists' exists in an area completely surrounded by that 'absolute nothing' and this means that 'that which exists' exists in a niche within the 'absolute nothing'. This 'niche' is the 'niche' of "The Niche Verse" (or "The Light Verse") in the Qur'an. This verse is as follows (the translation used is the translation provided by Abdullah Yusuf Ali):

(Surah 24:35): "Allah is the light of the heavens and the Earth. The parable of His light is as if there were a niche and within it, a lamp;

333

the lamp enclosed in glass; the glass, as it were a brilliant star; lit from a blessed olive tree; neither of the East or West, whose oil is well-near luminous, though fire scarce touched it: Light upon Light! Allah does guide whom He will to the light. Allah does set out parables for men; and Allah does know all things."

Now it seems to me that this verse is exactly describing the universe as I imagine it, yet my imaginings came before I read this translation of the verse. The niche is the area where God 'is'. This is the entire area that includes everything that exists including the ideas of the things that have existed and those that will never exist. Outside of that, there is truly nothing, so the something, if it could be viewed from the surrounding nothingness, would look like a niche and, within it an outer glass of anti-matter and an inner 'light' of matter - like the flame of a lamp shining within the glass. The entire thing, against the backdrop of nothingness, would appear like a brilliant star, a glowing 'something'. The olive tree is a reference to the 'Tree of Life' arrangement of dimensions that lights the lamp of matter with electro-magnetic radiation but the Tree is, in no way, East or West from anywhere. The 'oil' is the electro-magnetic radiation itself. Photons are luminous at many frequencies and fire doesn't seem to actually interact with light in a way that could be described as 'touching', as fire produces photons in both the visible and thermal areas of the spectrum.

What can be said of 'Light upon Light!'? Could it not be a way of expressing 'C^2'? It was the first thing of which I thought when I read the line. C^2 = 'Light upon Light!' The speed of light times the speed of light. It also describes the two main types of energy: matter and anti-matter, and how it appears from the outside looking through the slightly luminous 'glass' of anti-matter and the inner light from the matter within it: Light upon Light! First, the verse discusses that which exists, the matter and the anti-matter, glass and lamp; then, it discusses C^2. Put the two together - mass and C^2 - and you get Einstein's formula for energy: $E=MC^2$. So, it seems that this verse is the Qur'an stating Einstein's $E=MC^2$ equation in allegorical form!

It seems that, whilst Science has no answer to the question, 'Where did all the anti-matter go?' the Qur'an, does have an answer, although it takes a modern, Special Relativistic way of interpreting a verse of the Qur'an, which could only have been done since 1905. This is one of my main reasons for saying above that more modern interpretations can be drawn from the Qur'an and, therefore, should be, as they tend to support the fact that the Qur'an, indeed, had knowledge within it that no man, at the time, could possibly have had. If $E=MC^2$, then I strongly suspect that God knows it and knew it long before Einstein. The most knowledgeable entity in the universe should be able to describe relativity in a parable that is understandable by most and that would certainly draw attention when it was uncovered as being a statement, containing within it, thought that is perfectly in keeping with Special Relativity. There's every reason to do it, too, as it's another proof of the text.

Older languages didn't have words for 'matter' and 'anti-matter', but that doesn't mean you can't refer to them in some ways. A lamp within a glass is a very good analogy. 'Light upon Light' as being a pun that both imply the oppositely-charged kinds of mass and C^2 is a spectacular feat of language! In three short words, God has described the concept of C^2 and, in the same three words, implied the two types of mass (matter and anti-matter) and, thus, revealed the entire "$E=MC^2$" equation. This 'Niche Verse' or 'Light Verse' is not usually viewed as being one of the more prominent scientific statements contained in the Qur'an; but, I hope you will, now, see it as, perhaps, the most profound of them all.

The Parable of the Cup

The parable below was a creation of mine that I developed after my first reading of the Qur'an when I was about 17 years of age. The style of writing in the Qur'an and my own poetic bent made me think that I would like to try to write a parable in the kind of style that the Qur'an demonstrates. While Muslim readers may relate to this more, I still believe that the spiritual truths it contains will connect with most readers. So, here it is for your perusal:

When someone asks you, is the cup half full or half empty? Say to them: "Neither. It is the nature of a container to always be full. It is half full of liquid and half full of air and this is the truth of the matter." But, what is the cup and what are the liquid and the air? The cup is the human soul, for the soul is like a container. The water is sin, which seeks its own level. The air is Holy Spirit, which is the natural substance that fills a soul. When a soul acts immorally, it fills with sin and that sin will, if allowed, always seek to reach more sin and potentially fill the soul and make the cup of our soul overflow with sin so that the sin will spill out and reach the souls of others. However, when we act morally with God's guidance, Holy Spirit rushes in, of its own nature, to replace the sin and force it out. So we must act in such a way as to minimise the amount of the liquid sin in our souls by drinking it ourselves, in other words, accepting our own failings. In this way, Holy Spirit will charge in and fill us with that which is the natural substance of our souls, thus purifying our souls.

Final Word

Final Word

I sincerely hope that the reader reflects more upon the philosophical implications of this model of physics than the technical specifics of the model itself. It is purely through understanding the philosophical implications this model affords that it surpasses any previous models in its relevance for mankind. I may well have explained the answer to dark energy, where the Higgs boson is, and why time only moves forward in the process of writing this book; but I firmly realise that, whilst these discoveries have their importance for Science, they do little to shed light on the human condition. It is the human condition that I most desire to advance to the next level and that can only be done by the readers understanding the philosophy that can be drawn from this cosmology. Science tells us answers to 'how' but it doesn't touch the answers to 'why', as that has always been the realm of religion.

Finally, the two have a firm alliance. The answers to the 'whys' that spring forth from this model of physics give us all hope that, while we live, we have a chance - a golden opportunity - to raise our level of consciousness and that of those around us and make that leap forward that represents an evolution of our spirit. I believe that this type of spiritual evolution, based on an understanding of our universe and our role in it, is the key factor that will trigger the changes required to allow our brains themselves to evolve such that humanity will move on to become a new species - one that is more loving and compassionate towards our fellow humans and towards our environment.

Then, and only then, will we no longer be our own worst enemy. Then, and only then, will we no longer be the worst enemy our environment has ever seen. Then, and only then, can we face the future with our heads held high and be proud of humanity rather than hanging our heads in shame as we do now. I implore you to hearken unto this still, small voice crying from the wilderness to lead you back to that One who cherishes us all and, like that One, to cherish one another and to care for your fellow humans and our environment as best you can. Let this work make straight the path to that One who made us each a vital element of this creation and let us recognise that One as our greatest and only refuge. Let us get to know one another and work together in peace and not hate or murder one another for petty and transient causes. If we struggle, let it be that we struggle for peace and harmony and know that our differences are our greatest strength rather than letting them divide us. Then, and only then, can we share all the world, live life in peace, live as one and be as one. Please, do not just imagine that; please, 'for God's sake', do it!

New Lime

· PUBLISHING ·

We all know how serious and dangerous the issue of stress can be, but how many of us know what stress actually is, or how to efficiently deal with it? With the right education under your belt, it is a lot easier than you think!

Stress threatens more than our health ...It prevents us from ever reaching our full potential. With a little basic coaching in subjects such as time management, communication skills, motivation, attitude, accountability, and 'life balance', you can move forward and become more successful than ever before in the 'game of life'.

Don't just mask the symptoms of stress; ELIMINATE the source!

www.newlimepublishing.com
or available to order from any good bookstore
ISBN: 978-0-9571267-2-5

Lightning Source UK Ltd.
Milton Keynes UK
UKOW041026240712

196476UK00015B/9/P